FIREBALLS

FIREBALLS

A History of Meteors and other Atmospheric Phenomena

Craig R. Hipkins

To order additional copies of this book, contact:
Xlibris Corporation
1-888-795-4274
www.Xlibris.com
Orders@Xlibris.com
63502

CONTENTS

Acknowledgements 9
List of Illustrations 11
A note about the Illustrator 15
Introduction 17

Chapter 1: The Great Unknown 23
Chapter 2: Early Myths 28
Chapter 3: Ancient Fireball Sightings 41
Chapter 4: Fireballs in the Age of Chivalry 56
Chapter 5: Fireballs in the Age of Reason 74
Chapter 6: Fireballs in the Modern Era 96
Chapter 7: The Tunguska Fireball 106
Chapter 8: Siberia Again . . . The Sikhote-Alin Fireball 130
Chapter 9: Green Fireballs Over New Mexico 139
Chapter 10: Slow Moving Fireballs . . . The Comet 150
Chapter 11: The Lightning Fireball (Ball Lightning) 161
Chapter 12: The Will—o'-the Wisp 177
Chapter 13: The Fire of St. Elmo 191
Chapter 14: The Fireball of the Magi 203
Chapter 15: The Dragon Fireball 229

Bibliography (Books) 241
Atmospheric Phenomena Chronology 251

To my uncle Dale F. Wasilak (1949-1997)

Acknowledgements

I would like to thank my twin brother Jay S. Hipkins for his work in editing this book. Jay is a teacher at the Fletcher School in Charlotte, N.C. His skills in grammar were very helpful in presenting the final result of this work. He also gave me many useful suggestions over the past two years while I have been researching and writing this book. I would also like to thank my father Glenn A. Hipkins for providing the illustrations. I looked forward to coming home from work every day and driving over to his house for coffee to see what he had finished. Finally, I would like to thank my wife Rhonda, my mother Carolyn, and niece Paige for the encouragement they gave me while working on the book. My son Robbie also deserves to be mentioned. Many times I could hear my wife ask him what daddy was doing, and I would hear him reply "He's working on his fireball book again."

List of Illustrations

COURTESY OF GLENN A. HIPKINS

1. Fireball seen in the sky over Crete in 1478 B.C. (Chapter 2)

2. A Fireball breaks up over China in the year 616 B.C. It struck a number of chariots and killed 10 people. (Chapter 2)

3. A Fireball strikes the camp of the Roman general Pompey in 87 B.C. killing the general and many of his men. This fireball was recorded by many early historians such as Plutarch and Pliny the Elder. (Chapter 3)

4. A Fireball spooked the horse of the emperor Charlemagne while he was on campaign in Saxony in 810 A.D. (Chapter 4)

5. Gervase of Canterbury records that on June 23, 1178 a "burning flame" appeared in the eastern sky and "the moon was struck hard and slayed." In this illustration it is observed by some monks. (Chapter 4)

6. A great meteor shower observed over much of Europe in the year 1366 was described as "a movement of the stars such as men never before saw or heard of . . ." (Chapter 4)

7. On the evening of October 11, 1492 Christopher Columbus standing on the deck of the *Santa Maria* saw a light on the horizon

which has been described as "like a little wax candle bobbing up and down." (Chapter 5)

8. This illustration shows Mrs. Gardner of Wenham, Massachusetts observing a fireball streaking across the sky in the early morning hours of December 14, 1807. The fireball broke apart over the small town of Weston, Connecticut scattering debris over a large area. (Chapter 5)

9. The Tunguska fireball of June 30, 1908. This illustration shows the forest leveled in a radial pattern from the center of the blast. The trees in the center, although scorched and branchless stood like "telegraph poles." This was later determined to be caused by the fireball exploding in the atmosphere before striking the ground. (Chapter 7)

10. A Siberian fur trapper observes the great fireball of February 12, 1947 which broke apart over the Sikhote-Alin mountain range in eastern Siberia. (Chapter 8)

11. During the late 1940s and early 1950s the southwest United States was plagued by a rash of green fireball sightings. Some say they were natural phenomena, while others are convinced that they were caused by a secret government experiment or UFOs from another world. The cause of these fireballs has never been determined. (Chapter 9)

12. The Holy Grail was described as being "very bright" and "covered by a cloth of white samite." It entered King Arthur's great hall and hovered over the Round Table. It is quite possible that this was an early report of ball lightning. (Chapter 11)

13. The single combat fought during the search of the Holy Grail between Sir Bors and his brother Lionel was interrupted by a great ball of fire which fell between the two combatants. (Chapter 11)

14. The Will-o'-the Wisp is commonly seen around swamps or marshy areas as this illustration suggests. (Chapter 12)

15. The Fire of St. Elmo was often times seen on the masts of ships. Superstition regarded one fire as being the omen of a bad voyage, while two or more fires meant that the voyage would be successful. (Chapter 13)

16. A fireball lights up the sky over western Persia in the year 7 B.C. The biblical Star of the Magi has often been thought to be either a conjunction of the planets, a comet, or a supernova. However, it is quite possible that it was a very bright meteor. (Chapter 14)

17. In this illustration a fireball streaks across the sky of medieval Europe.

    Cover Art. This illustration shows the Assyrian army of King Sennacherib about to be destroyed by a fireball in the year 687 B.C.

A note about the Illustrator

Glenn A. Hipkins was born in Marlborough, Massachusetts in 1942. He worked for many years as a design engineer before retiring in 2002. He currently lives in Charlotte, North Carolina with his wife Carolyn. He has three children and two grandchildren. He enjoys drawing, and building model boats in his spare time.

Introduction

My interest in fireballs began back in 1984 when I was fifteen. At the time I was a sophomore in high school living with my parents and two brothers in Hubbardston, Massachusetts. Growing up during the era of Star Wars, I always had a keen interest in things related to the cosmos. The interest, however, was purely imaginative and not at all scholarly. A more emphatic and academic pursuit of the subject never crossed my mind, and indeed would not for many more years.

New England weather, as anyone who has lived in the region can tell you, can be extremely unpredictable. Our Puritan forbears on the Mayflower only took up residence there because a storm off the coast of New York forced them to turn around! To the best of my recollection the winter of 1984 was like any normal winter in New England. It was the spring of that year which was rather strange. A late snowstorm blanketed the region in early May forcing us to spend a day or two digging out of our humble little abode. Since school was closed, and ice fishing season was over, I was at a loss of how to spend this sequestered time. It was some time during those few days of this spring blizzard that I chanced upon a magazine article on the great Tunguska explosion of 1908. I have long since forgotten the magazine in which the article appeared, never mind the author of the piece. Unfortunately, a quarter of a century has clouded my memory. However, my memory possesses a vivid impression on what the article contained. It told of a fiery fireball that was witnessed by many people living in the Tunguska region of Siberia. The fireball flattened an area of tundra roughly equal in size to the state of Rhode Island. I was fascinated by the story. What could have caused such destruction? If I recall correctly, the article made mention of possible suspects, but the jury at that time was still out on the verdict. In the nearly twenty-five years since I read that article I have become more acquainted with the incident. Many competing theories vie for the answer. A meteor or rogue asteroid is the most popular, but a small comet, or mini black hole has their adherents. There is even a fantastical theory that it was caused by a crashed flying saucer. Whatever the cause, a century later the debate still rages on. I will examine the Tunguska fireball in more depth in chapter 7 of this book.

Although this article sparked a life long interest in astronomy, it wasn't until twenty-three years later that the event which led to the writing of this book came to pass. It occurred at around 7:50 p.m. on the evening of Wednesday January 24, 2007. Earlier that night my phone rang and the voice on the other end informed me of a railroad crossing signal in trouble in Harrisburg, North Carolina. The voice belonged to the railroad dispatcher, and the reason that he called me was because I am a railroad signal maintainer whose job is to take trouble calls when anything regarding a train signal or switch needs attention. There was nothing unusual about this night. It was bitterly cold, and I had the heater on in my truck listening to an oldies radio station. It was around a thirty minute drive to the crossing from my house, and when I arrived I found that another signal maintainer, thirty year veteran Bob Butts had also been dispatched to the scene. Reluctantly, I got out of my warm truck and met Bob in the signal bungalow where we carefully examined the readings on the motion detector that governs the crossing. The readings were erratic, and we decided that the problem was out on the track. We agreed that I would stay in the bungalow to observe the readings, and he would drive down to the other end of the circuit to see what he could find. He departed, and since the temperature outside was well below freezing I decided without hesitation to warm up inside my truck and wait for him to call me when he got in position.

It was a clear night, and my truck was facing north-east. I sat down and had barely shut the door when I found myself gazing at a bright green fireball moving slowly through the sky. My truck was on a slight incline so I had a clear view of the object through my windshield. The fireball was traveling in an arc about 45 degrees to the horizon. I observed it for about eight seconds which seems an eternity when a dazzling display like this takes center stage. The fireball was approximately the size of the full moon with an apparent magnitude of about—11. A persistent train streaked behind it somewhat lighter than the main body, and about 5 times its length. As I was watching it, a small piece of the fireball broke off and appeared to shoot upward. The fireball disappeared about 30 degrees from the horizon. There was no warning to its demise. It did not flicker out as a candle would do on a breezy night with your bedroom window open. It merely turned off as if someone had flipped a switch. I can best describe the event as a shocking display of brilliant light. My first reaction was "wow, what the hell was that thing." I sat in my truck peering through the windshield hoping that it would reappear knowing full well that it would not. Whatever it was had perished in dramatic style, that much was certain. One of the first things that came to mind was that it could have been some type of aircraft in a death plunge. I suddenly wished that I had been outside of my truck. Perhaps I could have heard a noise that may have accompanied it. Unfortunately I had been mesmerized by the visual aspect, and with the doors closed and the radio playing I had lost any chance to have been witness to its audible character.

For a few moments I contemplated what I had just witnessed. My initial fear that it was some sort of aircraft going down subsided once I had digested all of the facts. This green fireball had all of the characteristics associated with a meteor burning up in the Earth's atmosphere. The way it just turned off led me to this assumption. I have seen meteors do this before. Space offers very little resistance to a rock traveling 50,000 miles per hour. However, as soon as the rock enters the dense upper atmosphere it encounters resistance from air molecules that slow it down. Friction occurs which then causes the air around the rock to heat up producing the glow associated with the fireball. If the rock has enough mass it survives the trip through the Earth's atmosphere and hits the ground as a meteorite. However, most meteors do not survive the journey to the Earth's surface. They either become unstable and explode into fragments and become dust, or if the mass of the meteor is small to start with, it is merely ripped apart until there is nothing left. When the meteor explodes and leaves a visible vapor trail behind it, it is called a bolide. The fireball that I witnessed on the evening of January 24, 2007 broke into at least two pieces. After the lights went out I could see no vapor trail. However, had the fireball come during the daytime it is likely that this one would have left a smoky trail behind it which would probably have been visible for quite some time. Sort of like a vapor trail left by a jet.

Needless to say, I was excited. Never in my 38 years had I seen a green fireball. Over the last quarter of a century I have been an avid watcher of the night sky. I have familiarized myself with its many wonders. I have seen countless meteors, including a spectacular fireball during the Geminids meteor shower back in 1992. I had heard of green fireballs, but never thought that I would actually see one.

The next day I was surprised to hear that I wasn't the only one looking up at the northern sky the night before. Others had apparently witnessed the same thing. On Friday January 26, the front page of the Charlotte Observer carried an article about the fireball. The headline read "What Was That Light In The Sky?" below this "Unknown Freaky Orb drew eyes, imaginations across area." Mark Washburn, who was the author of the article, started out the piece with this sentence, "Earthlings, you've had an otherworldly experience, a cosmic encounter." He went on to describe eyewitness reports. A man in Blacksburg, South Carolina described it as "Bright blue-green ball with a white tail." A woman in Matthews, North Carolina said it was " A greenish-like light low in the sky." Another man in Gastonia, North Carolina described it as a "Large, bright green ball . . . had a haze about it."

WYFF in Greenville, South Carolina claimed "It is the talk of the town and beyond." "Reports came from as far away as Kentucky, Tennessee and North Carolina." Indeed, the strange green object that I had seen soon began attracting national attention. It found itself on the Drudge Report, and eventually made its way to the nationally syndicated late night talk show "A.M. Coast to Coast," a program that has a format that deals primarily with subjects like ghosts, aliens, and conspiracy theories. Websites around the globe picked up on it. Experts on aerial phenomena generally agreed that the strange orb like fireball seen over the Carolinas on the evening of January 24, 2007 was a meteor. According to the Charlotte Observer, Daniel Caton who is an astronomy professor at Appalachian State University, believes that it was a fireball that burned up about 30 miles high in the atmosphere. He might be right, but I keep an open mind.

I began to research fireballs with an earnestness that I had never before possessed. The more I delved into the subject, the more intriguing it became. Fireballs have been seen by man throughout recorded history. However, until just recently we knew very little about their nature or origin. Historically, they have been regarded with fear, superstition, and trepidation. In this book I will attempt to bring together mankind's experiences with not only the "otherworldly" visitors, but the ones created right here on our own planet. This is not a science book, and I am not a scientist. I am merely an observer and a gatherer of facts, stories, and legends. I am so to speak, a scribe, attempting to patch together a cosmic and sometimes almost ethereal experience that the living today have in common with our ancestors of yesterday.

Craig R. Hipkins
August 23, 2009

Chapter 1

THE GREAT UNKNOWN

People have been looking up at the sky for thousands, if not millions of years. Until just recently, the heavens, and everything contained therein were a big mystery. Humans attempted to explain this mystery using whatever primitive means they had at their disposal. The sky was the great unknown. It was a mystical place, an enigmatic place where the gods lived. It was a region sometimes revered, sometimes feared, but always beckoning to be understood. Even today when we look up at the sky, we remain awed by its majestic wonders and the vastness of its seemingly endless reach. Although we have a better understanding of it now than our ancestors did, we still do not completely comprehend the complex structure of what exists outside of the tiny planet we call home.

Primitive man had their own interpretation of the cosmos. In some civilizations the gods represented the planets and the Sun. The stars took the form of mythic beasts and cultish heroes who had performed deeds of valor on the terrestrial sphere, and were then cast into the night sky to be immortalized for eternity. The nomads of the Arabian peninsular believed that the sky was a blanket that covered the Earth at night to let it sleep. The stars were merely holes in the blanket.

It was not until the time of Copernicus in the 16th century that it became proven beyond all doubt that the Sun was at the center of our solar system. Until this time the prevailing theory was based on Ptolemy's Earth centered universe. It was a primitive and naïve way of looking at things. However, what were our forbears to do? There were no telescopes in those days, so people had to use what was available to them. In this case the naked eye was the best thing they had. It all changed when Galileo trained his refractor at Jupiter one clear night, and observed four satellites around it. They were, of course, Jupiter's largest moons. If Ptolemy had had this instrument at his disposal, he might have correlated the similarities between Earth and Jupiter. The Sun would then have been looked at in an altogether different light. Even so, the times were not ready for relegating Earth to the backseat of an impersonal and chaotic universe where Earth was, so to speak, "just one of the guys." Superstition and man's elated opinion of himself in the hierarchy of things stood in the way. When Copernicus introduced his heliocentric system in the 15th century, he was laughed at, and then when he did not back down, he was threatened with torture if he did not recant this abominable heresy.

Ptolemy's Earth centered universe fit better with the times. A quote from the 18th century British/American pamphleteer Thomas Paine sums up this close minded approach to science, invention and progress. He wrote "A long habit of not thinking a thing wrong gives it a superficial appearance of being right, and at first raises a formidable outcry in defense of custom, but the tumult soon subsides, time makes more converts than reason." In the case of the heliocentric system, that time was a long time in coming.

The ideas espoused by the Greek scholars of antiquity, and reaffirmed by the Romans in the first few centuries after Christ were as incontrovertible as Newton's law of gravity, and Kepler's motions of the planets are today. One of these early Greek scholars, Thales, was a mathematician who lived in the 7th century B.C. The only information known about him comes from the works of others. Some who lived centuries after his death. In fact none of his original works have survived, but his ideas have been recorded. At first they probably survived orally from teacher to pupil. It was only much later that they were finally written down. Thales believed that the stars were made up of the same material as the Earth, and that the Moon borrowed its light from the Sun. Most importantly he believed that the Earth was round.

If one examines these three axioms of Thales, it becomes quite apparent to the reader that if his ideas had been taken into serious account, the evolution of scientific thought would in all likelihood have evolved at a faster rate. It would have had at least taken a different path. Instead, most scholars, at least the influential ones of the day decided to disregard Thales reasoning, and instead follow the Aristotelian/Ptolemaic philosophy. In their universe, the Earth was fixed. It was a stationary object in which the rest of the universe was obedient and paid homage to its laws. In other words, everything revolved around the Earth, and was relative to mankind's being. It was the sovereign power . . . an immutable constant near the top of the hierarchal caste-like structure. Only the gods who resided in the heavens held a higher place than did Earth and its denizens.

One of the most common held beliefs of antiquity was that the sky was solid, and that the stars were stuck to this solid sheet like darts on a dartboard. They were lit every night and blown out like candles come morning. Plutarch writes of the philosopher Anaximenas who lived a century after Thales. Anaximenas believed that the sky was solid and crystalline and that the stars were fixed to its surface like studs. Although the ancients could easily explain the stars in this manner, the planets were a bit more difficult to explain. The stars were fixed, and when they changed position in the sky they changed en-masse. The planets, on the other hand were whimsical, and seemed to have no pattern to their movements. Sometimes, to the amazement of these early star gazers they moved in circles over the course of a few nights. Of course, we now know this seemingly backwards motion is nothing more than what is called an epicycle. It occurs when the Earth passes or is passed by a planet in its orbit around the Sun. The ancients, of course, had no way of knowing this. Instead, a common belief held that the planets merely moved along the solid background of the sky along another sphere.

So if the ancients thought they had figured out the nature and motions of the planets and stars, what must have been their opinion on the random objects that suddenly just appeared in the sky? Needless to say, their views on this subject varied as much as the riddle of how the universe formed does today. The common belief of the ancient world held that everything in the sky was held together by spheres which circled the Earth. These temporary apparitions were therefore looked at with a degree of fear. Since they were not a part of the permanent structure of the Earth centered universe, they could only have been cast out of the heavens by the angry gods.

Pliny the elder describes a "flame of bloody appearance" that fell down upon the Earth during the reign of King Phillip in the 3rd year of the 103rd Olympiad. Pliny stated that "nothing is more dreaded by mortals" referring to cosmic bodies which suddenly just appear in the sky. When analyzing Pliny's "flame of bloody appearance," we can see clearly that it was nothing more than a bolide. It was a cosmic event that Pliny and most people of his time did not quite understand, and would not for many centuries to come. Origen, a medieval Christian scholar believed that comets possessed souls and were living entities. Like many people of his time he also believed that comets portended some disaster, namely the downfall of a kingdom, or ill-luck in a battle. To people today it seems incredible that our ancestors held these views, but we are an evolving race. Even in this day and age there are people who hold wild, fantastical views of comets, meteors and other cosmic bodies. Only a mere decade ago, in 1997, Comet Hale-Bopp spelled the doom for a cult known as Heaven's Gate. The members of this cult believed that the comet was a sign that the end of the world was near. So they decided to commit suicide Jonestown style believing that their souls would ride the comet into another world. To them at least, like our ancestors before us, the sky was the big unknown.

We have a better understanding today of what we see in the sky. We are familiar with the nature of comets and meteors, the northern lights and other common aerial phenomena. However, there are still some things observed by the ever watchful and curious eye of the human that are not yet understood. For instance, speaking on a broad subject we bring up the problem of UFO's. What are they? Take a survey of 1000 people and you will probably get a thousand different answers. Are they spacecrafts from other worlds? Secret government experimental aircrafts? Or are they perhaps merely the wild fancies of a human populace that have fallen victim to a form of mass hysteria? The fact of the matter is that we do not know for certain what they are. So, if today we are still uncertain of things seen in the sky, one can only imagine Pliny the Elder in the 1st century gazing up at the heavens and describing something in the guise of "A flame of bloody appearance."

The knowledge of the sky has only recently been tamed to a point where we have begun to understand its complex structure. We have always strived to achieve the status of the birds and take to the sky. In a way we have managed to do this, and perhaps even surpassed these noble descendents of dinosaurs by being able to reach the limit of space and beyond. The desire to reach out into the heavens has been with us since the dawn of time. This can be seen in the mythologies and histories of the world. When a ball of fire is seen streaking across the sky, or an apparition of some unknown quantity suddenly appears in the heavens, the question naturally arises as to where it came from. In the next chapter we will study three of the earliest myths from world history to see just how our ancestors thought of these wonders of the sky.

Chapter 2

EARLY MYTHS

Manned flight has been a fascinating concept to humans since the dawn of recorded history. Mythology is filled with tales of flying chariots and men with wings that could fly with the birds and perform heroic and glorious deeds. At times, it seemed as if these men possessed supernatural qualities. One of the more famous of these early legends comes from Greek mythology and involves the story of Daedalus and his son Icarus. Daedalus was a brilliant inventor who had built a labyrinth for Minos, the king of Crete. The purpose of the great labyrinth was to keep the fierce Minotaur a prisoner within its walls. The Minotaur was a beast with a bull's head and a human's body. For one reason or another (the story varies) Daedalus had a falling out with Minos, and Minos had him confined within the labyrinth walls. It seems strange that the man who built this great piece of architecture could not himself find his way out, but indeed this seems to be the case. Although a prisoner of the king, Daedalus was given certain freedoms and was left with the tools of his trade. King Minos, it seems, was not about to lose the talented services of a man of Daedalus' ingenious and crafty ability.

Daedalus took advantage of the resources that were left at his disposal. He managed to make himself and his son Icarus (who was imprisoned with him) a set of wings which he intended to use to make his escape from the labyrinth. Before setting off he warned Icarus not to fly close to the sun, for the wax on his wings would melt. However, being a rash and brazen youth, Icarus was so enthralled with his new found freedom that he forgot his father's warning and flew so high that the wax on his wings soon melted. With his wings useless, he met his death with a fiery plunge into the sea.

The story of Daedalus and Icarus is a fable of sorts. It seems to send a warning for us to not get too caught up in the moment. The tale also alludes to the danger of too much freedom. People are never truly free. Freedom is fleeting and ambiguous, and it definitely has its limits as the tragic ending of the myth suggests. The story itself, however, takes on another meaning, a hidden one which we will examine shortly.

The Greeks were not the only ones who told of men that could fly. The legendary myth of Weland the blacksmith of Norse and Germanic tradition draws an eerie parallel with the Daedulus myth. Welund is sometimes portrayed in European folklore as being half man, half god, similar to The Iliad's hero Achilles. He is almost always known for his experience and skill in metallurgy. He is a blacksmith and armorer of the first degree. In fact, it is Welund's breastplate and chain mail which covers Beowulf in his legendary fight with the swamp monster Grendal.

In one story, Welund is taken captive by Nidud, king of Sweden. Nidud puts the blacksmith to work making armor and weapons for his army. Like Daedalus, he too is imprisoned in a great labyrinth. Welund, however, is a crafty smith and soon beguiles the king's two sons into entering the labyrinth where he murders them and fashions drinking vessels from their skulls and jewels from their eyes and teeth. He presents these ornaments as gifts to the king and queen who cheerfully and graciously accept them having no idea of their grisly origin. However, Welund's revenge is still not satisfied. The king's daughter Bodvild visits the labyrinth with a wish to have her ring repaired. Welund probably grinned as he concocted a potion which put the gullible princess to sleep, and lets him take advantage of her. He then decides that his revenge is satisfied, and like Daedalus, fashions a set of wings in which he uses to make his escape.

In another flying myth there is a Chinese legend which deals not with men using wings attached to their bodies but wings attached to flying machines. This myth is known as *The Land of the Flying Cart.* This enigmatic region of China is inhabited by people with only one arm and a three eyes. The third eye is set in the middle of the person's forehead. They travel around in carts that possess wings and can fly with the birds.

The three myths which I have mentioned are only a few of many which deal with mankind's early obsession with flight. These three tales show us that the sky and its mysteries were in our ancestors' thoughts early in our history. The reader may be asking what these fanciful tales have to do with the subject of this book. I can tell you that if we examine them a little bit more closely that it is possible each of them is directly related to a fireball. We must ask ourselves a simple question. What were the origins of these myths? Were they merely random stories that people thought up and decided that they were good tales to relate over a late night campfire? I do not think so. Joseph Campbell, the great author and mythologist famous for his book *The Hero with a Thousand Faces,* explained that myths were nothing more than metaphors.

At this late date along with the absence of the original record of the event, and probably hundreds if not thousands of years of oral tradition between the origin of the tale and its transcription on parchment, the true story will probably never be known for certain. However, by using Campbell's myth as a metaphor concept, a possible glimpse into their origin may be possible. The evidence relating to the three legends I have mentioned above as it relates to fireballs is tantalizing if not mind boggling. It is my aim now to show how this relation fits together.

I have already familiarized the reader with Minos the king who had Daedalus imprisoned. To understand the direction that I am going, a little history of Minos and the time he lived is in order. I might first begin by saying that when one talks of King Minos, one might be referring to any of the kings of the Minoan empire which flourished in the 2nd millennium before Christ. There is also a fine line between the Minos of myth and the Minos of reality. It is said in the Greek mythology that he was the son of Zeus and Europa. The date of King Minos' reign as related to the Daedalus myth is unclear, but it is believed to have taken place sometime during the height of Minoan culture between 1700 and 1400 B.C. A time known as the golden age of Minoan civilization.

Like many of the rulers of ancient times, King Minos was an opportunist. Daedalus had given the king what he wanted, but for Minos that was not enough. A man of Daedalus' caliber was hard to come by. The avaricious king could ill afford to lose him, so he had him arrested and imprisoned on a trumped up charge of assisting Theseus (slayer of the Minotaur) and the king's own daughter in their escape from Crete. As I have already related, the shrewd Daedalus made his escape. When Minos realized that his artful engineer had outwitted him, he took off in pursuit.

Daedalus had fled to Sicily and had taken up residence with King Cocalus of that land. It is said that Minos arrived in Sicily with his entire army but could not find Daedalus. However, he must have known his former prisoner well, for he knew just how he could flush him out of hiding. He set up a game in which the object was to pass a thread through the spiral of a snail's shell. Many people attempted the puzzle but none could figure it out until Daedalus appeared on the scene. Apparently impetuosity got the better of him. He was not the man to pass up an opportunity to show off the power of his brain. Taking an ant, he attached a thread on its leg and sent it through the shell. King Minos had Daedalus seized, but before the king could carry off his prize Daedalus had one more trick up his sleeve. He asked the daughters of King Cocalus to make up a bath for Minos. When the king was relaxed and caught off his guard, the daughters of Cocalus drowned him. Thus ended the saga of King Minos of Crete. At least this is how the legend ends. But what really happened to King Minos and the Minoan civilization? Popular conjecture holds that it underwent a foreign invasion of some kind, or that a volcanic eruption from Mt. Thera spelled its doom. Both of these theories are possible, but it is just as plausible that the Minoan culture was wiped out by a fireball. In fact, this theory holds great weight, and the reason that I believe it might be the answer is because there is a record of it!

In 1478 B.C. it is recorded that a thunderstone fell near the island of Crete. This event is recorded on the Parian Chronicle, and is one of the oldest instances of a fireball on record.

Illustration 1

Analyzing the myths of Daedalus and King Minos it can be vaguely seen that they possibly have a direct link to this so called thunderstone. Therefore, by using myth as metaphor we will now delve a little deeper into this mystery. We have already been introduced to the three main protagonists in this story: Daedalus, Icarus, and King Minos. Of the three, perhaps the least is known of Icarus, as he is only mentioned one time in all of the Greek myths. We are told that he is the son of Daedalus, and that he is a rather brazen youth that does not heed his father's warning and pays the ultimate price for it. But is that all we know about the son of Daedalus? I think not. Icarus is associated with the sun, and we know that a common belief of our ancestors held that meteors, comets and other stellar phenomena might come from the sun which of course, they believed, was powered by the gods. Was Icarus therefore nothing more than a metaphor for a fireball? Perhaps the one that was recorded by the Parian Chronicle as having struck off the island of Crete in 1478 B.C.? We know that Icarus plunged into the sea, and that this fireball was known to have hit the Earth in the sea off the coast of Crete.

We next find Daedalus fleeing to Sicily where he takes refuge with King Cocalus and his daughters. The king of Crete, however, is in hot pursuit, and has taken his whole army with him. This seems rather strange to me. Why take the whole army with him? Wouldn't a regiment or a battalion do just fine? Also, why has Minos himself gone in pursuit? Could he not have sent his most trusted General to accomplish this task? What is going on here? Why does it take everyone to take off in a passionate quest to capture one man?

History tells us that large armies are cumbersome and often become more of a hindrance to the commanding general. Logistics tells us this. A large army requires more food, more equipment, and more time traveling from place to place. Hannibal got along just fine with a smaller army. After all, he defeated the Romans three times, proving that big is not always the best. For Minos to take his whole army with him in order to arrest one fugitive is absolutely ridiculous. It would be the equivalent say of George H.W. Bush taking the whole United States army to Panama to arrest Manuel Noriega!

There is a simple explanation why Minos took the whole army with him to Sicily. It seems that they may not have been in pursuit of Daedalus at all. If they did finally catch up with him in Sicily, it was probably mere coincidence or luck which brought the King of Crete together again with his former engineer. The real reason was that the army was not in pursuit but in retreat! Indeed, it is possible that they were fleeing from the tidal waves caused by the fireball's impact with the sea. Also, we must remember the ignominious end of King Minos, drowned in a bathtub. Is this nothing more than a metaphor for the end of Minoan civilization? Were they victims of a tsunami? We can remember just recently the damage brought to the people of southeast Asia after the Boxing Day tsunami of 2004. This was later determined to have been caused by a slight shift in the oceanic plates. Tsunamis can cause catastrophes of major proportions. Did the island of Crete experience one following the fireball of 1478 B.C.? This fireball could very well have been one similar to the one that leveled the Tunguska region of Siberia in 1908 an event I discussed briefly in the introduction, and which I will later devote a whole chapter too. To be sure, this is only conjecture. However, it is not without merit. A Tunguska-like fireball would have destroyed the island of Crete. The survivors would have fled in terror to neighboring kingdoms. Witnesses to the fireball and the tsunami that followed would have spread their accounts throughout the land, and before long most of Europe would have heard of it. Possible theories would have been explored by the scientific minds of the day. For a few years the survivors would have told others of their experiences, including their children and grandchildren. Eventually, the survivors themselves would have passed into eternity. Their descendents would have passed the story down, but after a few centuries the event in its true form would have evolved into various myths and legends until eventually all that would be left of it would be the metaphor itself. Here then is the possible history of the legend of King Minos and Daedalus.

We now turn to the myth of Welund the blacksmith. This myth has a lot in common with the Daedalus myth. In fact, the two myths have so much in common that I believe the two men are one and the same. If we examine the evidence we can see that both Daedalus and Welund were made prisoners of selfish kings. They both constructed a great labyrinth for these kings. Both were master craftsmen of the highest order, and both men made their escape from the labyrinth by constructing a set of wings. From this evidence we can see that the two myths are at the least related. The Daedalus myth is the older of the two, and it seems plausible that the Norse or Germans molded the character of Welund from Daedalus. The fireball metaphor, though somewhat diluted in the Welund myth, is nevertheless still visible. However, if the Daedalus myth had not survived, it would be almost invisible. In fact, the only glimpse of the ancient fireball in the Norse myth comes when the crafty blacksmith makes his escape using a pair of Daedalus-like wings. There is no Icarus in the Welund myth, at least not in human form, but the Norse myth throws in a twist quite reminiscent of the Daedalus myth, but one really has to look for it.

Welund is a vengeful man. After escaping from the labyrinth, he flies directly to King Nidud's palace where he perches high up on one of the palace's great arches. King Nidud is enraged by Welund's impudence and calls in a great archer to take the bird-man down. However, he fails to take one thing into consideration. The archer's name was Egil who just happened to be Welund's own brother! By some prearranged agreement Egil was supposed to aim for a certain spot on one of Welund's wings. The spot was actually a bladder full of the two dead princes' blood. Egil had no trouble sending an arrow into the bladder, which burst sending the gore spilling onto the courtyard stones below. Welund's revenge was now complete.

In the Welund tale, the blood takes the place of Icarus and becomes the symbol of the fireball which fell over Crete over a thousand years earlier. As I have shown, this myth is so similar to the Greek myth that it is obvious that the one was derived from the other. The Norse have merely put their own spin on it in order to suit their own tastes in literature and oral tradition.

So, this leaves us with the last of our three metaphors: the Chinese legend of *The Land of the Flying Cart*. This is a rather mysterious and interesting myth. The origin of this tale is unknown, but like the two European myths that I have already related, we can catch a possible hint of it upon further reflection. The Land of the Flying Cart it was said, was adjacent to two other lands inhabited by persons that we would today consider to be prodigies in the manner of their strange physical characteristics. This enigmatic land is bordered by the Land of One Arm People. These people were said to have possessed three eyes, the largest one in the center of the forehead. To the southeast of the land of the flying cart was another land just as mysterious. The people who dwelled here were said to be Three Bodied People. Whatever is meant by this physical description is hard to say, but perhaps they were similar to the mythological hydra which were said to have seven heads attached to one body.

Not much is known about the Land of the Flying Cart. Even the geographical setting for this land is not known. The only other clue being that it was situated somewhere to the north of the Plain of Great Joy. This does not give us much information to go on. The Plain of Great Joy is similar to The Land of the Flying Cart in that the names themselves do not denote any specific geographical reference. The only clue in this respect would be the words "Plain" and "Land." This could refer to almost anywhere in China or Mongolia. "Land" encompasses the whole of China, while "Plain" though somewhat narrower in context, could refer to any flat region in that country. When one thinks of China, and a flat region, some sections of the Gobi desert automatically comes to mind. Is it possible that the Plain of Great Joy is just another name for the Gobi desert? It is hard to imagine anyone referring to a desert in such a hospitable light, but it cannot be ruled out. Wherever the Plain of Great Joy was, it appears that somewhere to the north of it in the year 616 B.C. a large fireball shot out of the heavens and struck a number of chariots killing ten people. According to the 19th century historian Daniel Kirkwood, this event is recorded in the Chinese Annals.

It would seem that this fireball must have broken up, possibly in the manner of the great Sikhote-Alin fall of 1947 (which I will discuss in detail in a later chapter of this book.) The fragments of this aerial beast may have fallen over a large area. Unfortunately we can only speculate on the rest of the story. The chariots must have been fairly close together.

A chariot race perhaps? In the flying cart myth the carts are said to be chasing each other through the sky instead of on the ground. The drivers of these carts have only one arm and converse freely and openly as if they are engaging in some kind of leisure activity. What can we make of all this? Is there a correlation between the fireball of 616 B.C. and the myth of the flying cart? There is no concrete evidence to show that there is, but like the Daedalus and Welund myths one can only wonder.

I have already shown that a real event over time can turn into myth within a few generations. If one needs further proof, one need only to look at the legends that have grown up about some famous people. What does George Washington have to do with a cherry tree? Or did Napoleon Bonaparte escape from the island of St Helene and live the rest of his life in a humble state in some foreign land under a rogue alias? Probably not, but who knows for sure now? Reality and myth have become intertwined so that no matter how much evidence is given to negate the myth, there will always be some that will believe it.

In regards to the Land of the Flying Cart I present to you this possible scenario. A group of chariots are lined up along a dirt track at an outdoor arena similar to the scene in the classic Charlton Heston film *Ben Hur.* There are many anxious people gathered around getting ready to watch this event, possibly the Super Bowl of its day. Suddenly the sky lights up, and the spectators shift their gaze from the stalwart young heroes on their chariots to a glowing fireball streaking at an odd angle across the sky. There is a hushed silence as they stare up in awe and wonderment. However, the silence is quickly broken when suddenly there is a deafening roar, and the spectators amazement is replaced by fear and horror as they are pelted by a hail of rocks and fragments as the fireball breaks up. Many people are killed and wounded including ten of the heroes who were so preoccupied with the drama on the track that they never saw it coming. News of this tragedy spreads far and wide, and eventually some poet or scribe immortalizes the event. Over the centuries, the event is retold and eventually the chariots become carts. Since the means of death came from the sky, it is there that the heroes should spend a happy eternity, maimed and disfigured though they might be.

Illustration 2

Admittedly, this analogy that I have come up with would be hard, if not impossible to prove. The chariots may not have been at an arena at all. They could just as well have been preparing for battle, or part of some trade caravan. The details of this fireball are few, and sketchy at that. I have concocted this possible scenario only to show that fireballs in our ancient past could be related to some of the myths that have managed to survive up until our time.

It seems incredible and rather far fetched for a scenario like this to happen, but if you are having a hard time believing that it could, try not to be too cynical. Ten men killed by an exploding aerolite may be a rare event indeed, but it cannot be totally dismissed out of hand. In 1674 it is recorded that two Swedish sailors were killed on the deck of their ship by a rock falling from the sky. We can also ask the question "What if?" On April 26, 1803 one of the most talked about fireballs ever recorded is known to have exploded over the small farming community of L'Aigle, France. The fireball was accompanied by a loud roar which some witnesses said sounded like thunder. This noise could be heard in places up to seventy miles away. After scouring the countryside, more than 3000 fragments of this meteorite were found. This is an extraordinary number, and one can only imagine the horror and destruction this fireball would have caused if it had struck a heavily populated community. I will discuss the L'Aigle fireball in more depth in a later chapter of this book.

The three myths that I have discussed in this chapter are tales that have been told and retold over the centuries. Each telling may introduce a new detail to the story. It can be safely said that these myths have evolved over time. It is also quite apparent that each different culture has borrowed myths from their neighbors and used them to suit their own purpose. In this chapter I have discussed only three instances where a myth has evolved from a possible encounter with a fireball. There are many more stories to be told, but for now we will leave the realm of flying men and machines and head into a region in which we will encounter what it was like to view a fireball in the ancient world.

Chapter 3

ANCIENT FIREBALL SIGHTINGS

Fireballs sightings have been recorded since human beings began to express themselves in art and writing. An African tribe called the Dogon who live in what is now the country of Mali recorded on cave walls images of stellar objects. Some of them appear to resemble comets or meteors. These drawings are thought to be thousands of years old. It is possible that they are the oldest depictions of a fireball that exist today.

In the mid 19th century an Englishman by the name of Austen Henry Layard was traveling through the Middle East and happened to stumble upon some earthen mounds in the ancient city of Nineveh in what is now Iraq. The mounds were dated to the third millennium B.C., which was during the height of the Assyrian empire. Found amongst all of the debris were some broken clay tablets. There was a form of pictographic writing on them which is called Cuneiform, and is similar in style and form to Egyptian hieroglyphics. The tablets were translated and found to contain an early history of the Sumerian-Assyrian culture. The most interesting part of the translation told the story of a legendary Sumerian king named Gilgamesh. Gilgamesh, king of Uruk was the archetype of the hero. When it came to prowess and strength, he can best be compared to the Celtic warrior Cuchulainn who is the hero of the Irish epic *Tain Bo Cuailnge* (Cattle Raid of Cooley). Both stories involve possible encounters with fireballs.

The story of Gilgamesh is probably based on a real king of Sumeria who lived sometime around 2700 B.C. Undoubtedly, part of the tale is myth, but as the old cliché goes—for every myth there is a shade of truth attached to it. Before Gilgamesh sets out on his adventure, he has a dream that a meteor falls from the heavens. Gilgamesh attempts to lift the stone, but it is too heavy. Only with the help of his mother, the goddess Ninsun, is he able to lift it. The stone is a metaphor, supposed to represent "Enkidu" who becomes the companion of Gilgamesh and shares the main protagonists peregrinations and adventures.

Turning to the *Tain* we find the hero Cuchulainn engaged in a war with the king and queen of Connacht over the removal of a great brown bull. It is unclear exactly when the "Cattle Raid of Cooley" took place, but it was probably sometime around the 1st or 2nd century A.D. This was well before the introduction of Christianity in Ireland by St. Patrick in the 5th century. Shortly before the great battle in which the conflict is finally resolved, the Connacht king, whose name was Ailill, sent a warrior named MacRoth in front of the army to scout the enemy position. MacRoth returned and gave a strange report to his king. He claimed to have seen "a dense fog which filled the valleys and hollows." Through this fog he could make out "sparks of fire of all different colors," and "flashes of Ailill sends a warrior lightning, with a great uproar and thunder." He then reported that there was only a slight wind that day, but as he was standing on the hill looking at the plain before him "a great wind came that flung him on his back, and almost swept the hair from his head."

Upon a first look at this strange report it appears that MacRoth may be observing the first stages of the battle, but it must be remembered that he was sent out to scout the enemy in front of his army. Of course it is possible that he might also be reporting on a lightning storm. However, it would have to have been an unusual one to be throwing off sparks of fire of all different colors. Could this have been a report of a fireball? It seems logical to me. The great wind that flung him on his back is reminiscent of some eyewitness accounts of the Tunguska fireball of 1908 which was said to have had a wind similar to the one in this ancient report.

Leaving *The Tain* I now venture off into the age of Homer, for I would be neglectful if I did not mention a certain incident that took place in *The Odyssey* that possibly involved an encounter with a fireball. Zeus, the son of Cronus, and supreme ruler of the Greek pantheon of gods, was well known for letting loose his thunderbolts when something displeased him. For the most part, it was thought by the Greeks as well as by other ancient cultures that lightning and thunder were caused by the anger of the gods. So, when in classical Greek literature Zeus is seen casting his thunderbolts down to the Earth, it is only logical to assume that an electrical storm was due to his rage. However, sometimes fire and brimstone are introduced into the mix, which gives us pause for thought. In book 12 of *The Odyssey,* the hero Odysseus and his men have funneled the ire of Zeus by butchering a herd of prized cattle on the island of Sicily which had belonged to the god of the sun. Leaving the island, Odysseus and his warriors set sail in their ship. However, it isn't long before they are caught in a terrible tempest. Needless to say, the storm contains fire and brimstone which tears apart the vessel drowning everyone on board except for Odysseus who manages to ride the waves all night and eventually washes up on a lonely outcrop of rock in the morning.

It must be remembered of course, that *The Odyssey* is considered a work of fiction. It was a poem written by an unknown poet who history has assigned the name "Homer." It is often said that he lived sometime around 800 B.C., but absolutely nothing is known of the writer's life. If *The Odyssey* is a fable of sorts, there is probably some vestige of an actual event in the poem. Up until the late 19th century it was thought that the city of Troy was only mythological. However, in the late 1860's a German-American businessman named Heinrich Schliemann found the legendary city while excavating in what is now western Turkey. This proves that at least part of Homer's other work *The Iliad* is true. So now if we look at the destruction of Odysseus' ship in this light, we can see that it is possible that at least some of the story is based on an actual account.

Removing ourselves now from the realm of history's earliest heroes we turn to the historians themselves who recorded some of the earliest fireball sightings. The Parian Chronicle whose author or authors are unknown, records a fireball that appeared over Crete in the year 1478 B.C. As I have already mentioned this fireball in the last chapter, and associated it with the collapse of the Minoan empire, I need not expound on it here except to say that this was not the only fireball seen and recorded over the island of Crete in the 2nd millennium B.C. The chronicle also mentions a fireball falling out of the sky over Mt. Ida in Crete in the year 1168 B.C.

The Bible has its share of fireball sightings. In the Old Testament's book of Joshua we are told of a great shower of stones hurled out of the heavens at the army of the Amorite kings whom Joshua was fighting. The passages are somewhat vague and ambiguous, for it is not clear if the rocks from the sky are hailstones or the debris of an exploding meteor.

Joshua 10:11: "While they fled before Israel along the descent from Beth-Horon, the Lord hurled great stones from the sky above them all the way to Azekah, killing many. More died from these hailstones than the Israelites slew with the sword."

Normally people do not die from hailstones. There are, to be sure, rare cases where death has occurred as a result of hail. Randy Cerveny, in his interesting book on strange weather phenomenon, *Freaks of the Storm,* mentions an incident that took place in India in 1888 in which a hailstorm killed 246 people. It is said that the hailstones were as large as goose eggs and cricket balls. Could this have been an exaggerated number? Possibly, but if a large group of people were caught out in the open, unprepared, without any cover, it is quite possible that a catastrophe like this could happen. We know that traditionally, armies are expediently forced to quarter themselves outdoors. Throughout history great armies have been forced to endure cruel and adverse weather conditions. One only has to remember Washington's army at Valley Forge, or Napoleon's retreat from Moscow to find two classic examples. An army being pelted by hailstones, or the fragments of a meteor would be just one more example of a encountered hardship.

The Roman historian Pliny the Elder mentions a fireball which took place during a show of gladiators in the 1st century A.D. Pliny was an astute and dedicated chronicler of natural events. Prodigies of the sky seemed to be of high interest to him, even if he did not understand them that well. He mentions a fireball that was seen during the consulship of Valerius and Marius. It was described as being a "burning shield which darted across the sky from west to east at sunset." This fireball threw out sparks in all directions. Based on this report, it appears that this "burning shield" was most likely a meteor that exploded in the atmosphere.

From the reports that Pliny has passed down to us, it seems that he must have had access to a good library. He mentions the Greek philosopher Anaxagoras, who is perhaps best known for suggesting the world was made up of tiny particles called atoms. Anaxagoras, however, was also prone to make predictions and prophesies of things to come. He foretold of a great stone falling from the sun and striking the earth at a certain time. This was a bold prediction, but according to Pliny he made good on it. In or around the year 467 B.C. a fireball or a comet was seen in the sky above Thrace on the Greek mainland. The fireball was seen in the daytime so it had to have been an amazing spectacle to anyone who witnessed it. Shortly thereafter, a stone as big as a wagon or a cart was found near the river Egos. The stone had a "burnt appearance," undoubtedly a stony meteorite. Pliny added that there was some talk on whether or not the stone had fallen from the sun. One can easily imagine the talk and confusion that this sensational event must have caused.

Once word got around that the prophesy of Anaxagoras had been fulfilled, he must have been looked at as a seer of utmost importance. If anything came from this fireball, it was the notion that rocks could fall from the sky. Pliny's theory of the cosmos was primitive, and fit in with the age in which he lived. He did, however, realize that rocks fell from the sky, so it is kind of hard to believe that scientists and scholars of the late 18th and early 19th centuries had a hard time believing this until it was proven to them by the falls at L'Aigle, France and Weston Connecticut in 1803 and 1807 respectively.

One more thing needs to be brought up regarding Anaxagoras' fireball. The fireball itself could very well have been a comet, and the stone of "burnt appearance" a meteorite that just so happened to coincide with it. Pliny often gets the two stellar phenomena mixed up. As can be seen in his description of "an ominous appearance in the heavens, that was seen once only." He described this as "a spark falling from a star, and increasing as it approached the earth, until it became of the size of the moon, shining as through a cloud." This almost certainly is a description of a bolide. However, a bit of confusion arises when he then adds "it afterwards returned into the heavens and was converted into a lampas." The word "lampas" is Greek, and when translated into English means "torch." "Lampas comes from the root "Lampein," which means "to shine." So we see here that Pliny was describing something bright, and most probably this was a comet. Did Pliny believe that meteors could turn into comets? It would seem so from this evidence.

We now take our leave of Pliny and examine the works of another Roman historian of antiquity, Julius Obsequens. Obsequens was a chronicler of sorts whose only interest seemed to be prodigies or strange events. He probably would have been known as the Charles Fort or Robert Ripley of his day. He is often associated with Livy, and indeed seemed to borrow heavily from his works. Nothing is known of his life except that he probably lived in the 4th century A.D. Some of the anecdotes that he has left us are interesting, and even entertaining. At the same time, some of them are even ludicrous. For instance, in the year 104 B.C. he tells us that "In Lucania two lambs were born with horses feet; one of them had the head of a monkey." It sounds to me like this was someone's idea of a practical joke, (an early Foy's ape) or perhaps some pagan ritual that has long since been lost from memory. Another prodigy that was supposed to have been observed took place in the year 134 B.C. and is just as absurd and humorous. Obsequens says that "On the capitol at night a bird uttered groans which sounded human." This book is filled with such crazy anecdotes, but it is also chock full of anomalies seen in the sky. Some of them, obviously, early eyewitness reports of fireballs.

For the year 167 B.C. Obsequens records; "At Lanvium a blazing meteor was seen in the sky." Another one from 137 B.C. is a little more descriptive, and is most certainly a bolide "At Praeneste a blazing meteor appeared in the sky, and there was thunder from cloudless heavens." This is one of the earliest descriptions we have of a fireball accompanied by noise. For the year 106 B.C. he records that "An uproar in the sky was heard, and javelins seemed to fall from heaven." In the year 91 B.C. he states that "While the war with Italy was gathering during the legislative activity of Livius Drusus, tribune of the commons, many portents appeared in Rome. About sunrise a ball of fire flashed forth from the northern heavens with a great noise in the sky." As is common with early reports of comets and meteors, Obsequens seems to correlate this fireball with the unstable political atmosphere of the day. One of the strangest reports recorded by Julius Obsequens was one that he chronicled for the year 87 B.C. He states: "While Cinna and Marius were displaying a cruel rage in their conduct of the Civil War, at Rome in the camp of Gnaeus Pompeius the sky seemed to fall, weapons and standards were hit, and soldiers struck dead. Pompeius himself perished by the blast of a heavenly body."

Was this a hailstorm accompanied by thunder and lightning? Or was it a barrage of stones falling from the sky? Perhaps an early Sikhote-Alin type fireball that just happened to explode over Pompey's army. "The blast of a heavenly body" seems to support this, for "a body" would seem to imply that it was something solid. Our old friend Pliny also made note of this fireball. He says that Pompey was "paralyzed by a star." This is a rather bizarre thing to say. However, when taking into consideration the limited knowledge that the ancients had of the stars, it does not seem so unbelievable. Is it possible that Pliny was alluding to a meteor? We have already noted that he was aware of rocks falling from the sky through his study of Anaxagoras. So it just might be possible that this is what he meant. We have also noted that there are very few instances where people have been killed by hailstones.

Illustration 3

Plutarch, in his life of Pompey, also mentions the incident. In order to diffuse any confusion that the reader might have it is necessary to clarify one thing. The Pompey killed by the heavenly body is not the same Pompey that Plutarch has honored with a biography. Indeed, he is the father of that famous general who would go on to become a member of Rome's First Triumvirate along with Crassus and Julius Caesar. The elder Pompey, writes Plutarch, was "killed by a stroke of thunder." We can infer one thing from the three sources that describe the death of Gnaeus Pompeius. He died a violent death, and although it is quite possible that he was killed by lightning or hail, we cannot rule out the possibility that a fireball from space was the culprit. Let us say that the jury is still out on this one and probably will be for a very long time.

We now move on to one of history's most enduring mysteries. This would be the emergence of the Phrygian goddess Cybele in Rome in the late 3rd century B.C. The history of Cybele is ancient, possibly dating back to the time of the Hittite civilization fifteen centuries before Christ. Cybele was the "Magna Mater" or "Great Mother" of the Phrygian people, and was thought to have resided in a temple on Mount Ida in Anatolia. There are various tales which have grown up around this goddess, but one of the more popular and enduring stories involves Cybele's lover Attis, who, after a quarrel with the goddess, goes insane and castrates himself under a pine tree. The Phrygians commemorated this event at an annual festival. When the Cult of Cybele was brought to Rome in 204 B.C., the Phrygian priests continued to celebrate it by cutting down a pine tree and carrying it to the Palatine where it is said that Cybele herself was personified by a black rock. It is told that the pine tree was garlanded with violets which supposedly were used to represent the body of Attis.

So what does the Cult of Cybele have to do with a fireball? It is the black rock which was carried back to Rome by the Roman General Scipio Africanus that interests us. The stone had long been revered in the city of Pessinus in Anatolia. Legend has it that it was there that the stone fell from the bosom of Cybele. Interestingly, the city of Pessinus is where the legendary King Midas is thought to have hailed from, and indeed he is often said to have been the son of Cybele.

The question now arises as to why Scipio thought it was necessary to remove the rock from Pessinus and bring it to Rome? Also, what was so special about this rock that the Romans felt that it needed to be located in their capitol city? Supposedly it represented the goddess Cybele, but why was Scipio so interested in an Asian/Greek goddess? The answer lies in the garb of superstition. The Romans were always looking for an edge, and in some respects were resourceful and pragmatic in their way of attaining it. The year 204 B.C. was an important year in the history of the Roman Empire. Hannibal, the great Carthaginian General was once more threatening Rome. A few years earlier he had dealt the Romans two crushing blows in battles at Lake Trasimene and Cannae, victories of strategic brilliance which solidified his reputation as one of history's greatest generals.

The Romans were at a loss as to what to do about the Carthaginian problem. It was decided that the Sibylline Books needed to be examined in order to see if or what could be done to solve it. The Sybles were female prophets who it was said could read the future. The king of Pergamum at this time was a Roman ally by the name of Attalus. A Roman delegation was sent to Pergamum to consult with Attalus, and determine the best course of action to be taken. The books were examined, and it was concluded that the only way that Hannibal could be defeated was if the stone of Cybele was removed from its palace in Pessinus and brought to Rome. Attalus, who was a Roman sycophant, wholeheartedly supported this plan, and it was therefore carried out. The black stone of Cybele was transferred from Pessinus to Rome with great pomp and circumstance. Less than two years later, the great Hannibal was forced to flee Italy for Carthage, and was then soundly defeated by Scipio at the battle of Zama in 202 B.C. Thereafter, the stone of Cybele was consulted by the Romans on many occasions, even if it did go against the core of their polytheistic religion.

So what was the origin of this black stone which was held in great reverence and veneration by the Asians, Greeks, and Romans alike? There are a few possibilities, but it is not known for sure where the stone came from, or when it actually fell from the sky. It is quite possible that it could have been a fragment of the great fireball of 1478 B.C. which I talked about in the last chapter. However, I am of the opinion that this fireball struck the ocean, so it would appear unlikely that it would be related to the stone of Cybele. Earlier in this chapter I mentioned another fireball which fell somewhere near Mt. Ida in Crete in the year 1168 B.C. It is interesting to note that there are two mountains by the name of Ida. One of them is on the island of Crete, and if you have not guessed it already, the other one is located in Turkey! Is it possible that the Phrygian Ida is named after the mountain in Crete? It is food for thought and leaves us open to further speculation.

Another fireball mentioned by numerous writers is one that fell in the year 465 B.C. in Thebes. It is said that the Greek poet Pindar was sitting on a hill when the fireball blazed through the sky and crashed into the ground at his feet. According to the 19th century astronomer Daniel Kirkwood, the stone was encircled by fire as it fell, and was said to be of "moderate dimensions, of a black hue, of an irregular, angular shape, and of a metallic aspect." Parts of this story are almost certainly apocryphal. However, by the description of the stone it would seem that it was definitely a meteorite. Some writers have referred to this stone as the one worshipped at Pessinus, and later at Rome. It is also said that this stone could cure madness and purify the body.

Whatever the origin was of the stone of Cybele, the cult that worshipped the "Magna Mater" was an enduring one. In Rome it lasted at least up until the time when Christianity had trumped all other religions during the reign of Constantine in the 4th century. Indeed, vestiges of this cult can still be seen today in the rebirths of pagan religions around the world. Some of them are pantheistic in nature, personifying the Earth itself as a female deity or goddess which is often used as the core or central belief of the religion.

Taking our leave of the stone of Cybele, we now turn our attention to another part of the world. Thus far, most of this chapter has been devoted to the region which lies on or near the Mediterranean Sea. However, Rome, Greece, and Turkey were not the only areas where fireballs were recorded in ancient times. The countries of China and Japan also had their fair share of sightings. In 644 B.C. fire stones fell from the heavens in China at a place called the "Country of Song." In the year 211 B.C. there is a record of multiple stones falling in China accompanied by a falling star. In 89 A.D. it is told that two large stones fell from the sky at Yong in China. It was said that the sound of this fireball was "heard over forty leagues." T.L. Phipson a 19th century British scientist made mention of a fireball from Mongolian folklore. According to this legend a black fragment of rock forty feet high fell from the sky on a plain somewhere near the source of the Yellow River in western China. He said that the Mongolians called this massive rock "Khadasut Filao," which translated into English means "Rock of the Pole." The date of this event is unknown, but it is undoubtedly of ancient origin.

At a Shinto shrine in Nogata Japan is a meteorite which dates back to 861 A.D. It has the distinction of being the oldest known witnessed fall of a fireball in which the stone still exists. For many years it was thought that the great fireball of 1492 held this honor. This fireball produced the stone at Ensisheim in France. I will discuss this famous sighting in more detail in the next chapter.

The final thoughts that I would like to mention in this chapter on ancient fireballs has to do with a possible record of one in the early Hindu scripture known as the Rig-Veda. The Rig-Veda is a collection of hymns originally written in Sanskrit sometime around 1000 B.C. The chief god of the ancient Hindu faith is Indra. Indra could control the sun, and had power over the animals and water. He can best be compared to the Greek god Zeus or the Egyptian deity Ra, and like Zeus, Indra was fond of using the thunderbolt to wield his power. He is often associated with a powerful drink called Soma. It is unclear exactly what Soma was, but the effects that it had on Indra were extreme, often causing him to engage in destructive behavior. This is especially true when he used it in the presence of his enemies. It is quite possible that Soma was some kind of hallucinogenic drug. It could also have been a potent alcoholic beverage of some kind, perhaps laced with some other drug that made the user feel a sense of omnipotence.

The author of this book experienced the affects of a powerful drink while serving in the Marines on the island of Okinawa back in 1987. Here, the generic term for a punch-type alcoholic beverage was called Mo-Jo. Every bar had its own version or recipe. There was on one occasion, an instance when the author got a little bit more than he bargained for after imbibing a large quantity of the house special. I have long since forgotten the name of the bar, but I can remember some details of its structure and its furnishings. It was located on the second floor of a house. It was a rather spacious place with three long oval tables which were equipped with bench seats that had been bolted to the floor. The place was by no means meant for comfort, for there were no booths or even chairs with backs on them. The entertainment consisted of a large screen television, years before they became common in households. It was built high into the wall in a central part of the bar where it was visible to everyone seated on one of the benches. The peculiar part of this set up was that there was no sound coming from the television itself. The sound came from a different source behind the bar, where a stereo played heavy metal music which blared over the various speakers placed strategically all over the room. The television itself served only one purpose, and that was to bring out the beast from within the person viewing it. On the screen were scenes of destruction and death— cars exploding in flames, airplanes crashing, and people being beaten or executed by firing squads. These were just some of the horrific images which came across the screen. This violence, coupled with the almost satanic grinding of the loud music had a noticeable effect on the bar's patrons, especially after the house drink had gotten hold of them. The special that night was a type of red Mo-Jo served in a carafe which I slowly sipped with a few of my buddies while we became transfixed with the images on the big screen. There were at least fifty other people, men and women in the bar that night— most of them drinking the same beverage. After consuming a good quantity of the Mo-Jo, I felt myself drifting off into a sort of ethereal world. I began to hallucinate and see things that could not possibly exist. A cockroach on the restroom floor appeared to be the size of a house cat. As the night went on, the affect of the drink, and the corrupted atmosphere produced the affect that it was meant to attain. The general hypnotic feeling that at first encompassed the whole room had evolved into an intense atmosphere of rage and paranoia. Fights broke out, and the savage and primitive nature of man took hold of even the most benign individuals. The place had turned into a lunatic asylum of sorts, and although I had consumed a good quantity of the drink, I still had the good sense to remove myself from this environment before something regrettable happened. I recall staggering out the door, and having to conquer a set of stairs which led to the street below, and had been painted blood red for a reason!

Anyway, I mention this incident only to show that the Soma of Indra could have been something similar to the Mo-Jo I drank on Okinawa over twenty years ago. The Rig-Veda tells the tale of the symbiotic relationship of Indra and Soma, and their battle against the Rakshasa which was a beastly creature who was described as having cannibalistic tendencies in the fashion of Grendal in the tale of Beowulf. Indra conjoined with Soma are a formidable force to tangle with. A humorous way of comparing the chief Hindu god and his favorite beverage would be to liken it to Popeye and his spinach!

In the mid 19th century the Oxford scholar H.H. Wilson translated the Rig-Veda into English from the original Sanskrit. The translation is complete, but much can be said of the fluidity of the hymns which can be somewhat confusing. However, one thing stands crystal clear from the text, and that is the complete destruction of the Rakshasa from the combined powers of Indra and Soma. The first verse of the 5th Ashtaka, 7th Adhyaya, 15th Sukta states "Indra and Soma, afflict, destroy the Rakshasa;" In the 3rd verse Indra and Soma "chastise the malignant Rakshasa, having plunged them in surrounding and inextricable darkness, so that not one of them may again issue from it." However, it is the 4th and 5th verses of this Sukta that garners our interest here. The 4th verse states "Indra and Soma, display from heaven your fatal weapon, the extirpation from earth of the malignant Rakshasa, put forth from the clouds the consuming thunderbolt, wherewith you slay the increasing Rakshasa race." At first glance this fatal weapon would seem to indicate a bolt of lightning. If you stopped reading here that would be the general impression that you, the reader, would get. However, all one has to do is jump into the next verse to see that this isn't so. It says "Indra and Soma, scatter around your weapons from the sky, pierce their sides with fiery scorching adamantine weapons, so that they may depart without a sound." This is an interesting statement, and there are a few clues within this verse that unveils its true meaning. Scattering the weapons from the sky would seem to imply a bombardment of something. Could this mean lightning bolts or hail? Certainly, but the words which come directly after this seem to annul this possibility. They read "pierce their sides with fiery scorching adamantine weapons." Obviously from this, we can gather that the weapons that Indra uses are terribly hot. The words "fiery" and "scorching" tell us this, but what about the word "adamantine?" What does this mean? The word "adamantine" is a derivative of the word "adamant" which Bantam-Scribner defines as "extremely hard or unyielding," "unconquerable," or "a real or imaginary stone of great hardness!" As the old Yukon prospectors would say . . . It seems that we have hit paydirt!" In fact there is a mineral named adamantine (a form of corundrum), which has a rating on the Moh's Hardness Scale of 9. Only a diamond with a rating of 10 ranks higher. So what does this all mean? Wilson leaves a footnote after this verse which he notes rather succinctly "The text, after scorching, adds another Epithet, Ajarebhih, ageless, undecaying." The sanskrit word "Ajarebhih" is rather obscure, but the few references that I have found for it allude to the same meaning. The 19th century Englishman, John Muir (not the noted American naturalist), who also translated the Rig-Veda as well as other ancient Sanskrit works, refers to the word, and also translates it as "undecaying." This word adds fuel to our fire. Is the Rig-Veda telling us something about the make-up and composition of a fireball? The term is said to appear after the translated word for scorching appears. The "adamantine" or "hardstone" whichever you prefer to call it had been moving through the vast reaches of space for eons. Did the author or authors of the Rig-Veda have knowledge of this? If they did, this is just one more example of the ancients possessing a knowledge that had not been tapped by later scholars until the last two centuries. It is interesting to ruminate this possibility. So, this having been said, I now close this chapter and venture into the age of medieval fireballs.

Chapter 4

FIREBALLS IN THE AGE OF CHIVALRY

In the last chapter we talked about fireballs recorded in the days of our planets' earliest civilizations up until the time of the Roman Empire's demise. The better part of mankind's time on earth has been spent in a period of uncivilized barbarism. We now enter a period in our history which for a lack of a better term is often referred to as the "Dark Ages." This term is really a specious one. It insinuates that the period of time following the collapse of the Roman Empire up until the time of the renaissance was one of uncultured ignorance. In actuality this term is a misnomer. Nothing can be further from the truth. An abundance of literature has survived from this period. One only has to look at the great works of Gregory of Tours, Geoffrey of Monmouth or the Venerable Bede to see this. All three of these writers will be examined in this chapter. This period also produced the world's first illuminated manuscripts like the Irish *Book of Kells*. Monasteries filled with studious monks translated and copied ancient manuscripts, and at the same time created new ones. Most of these works were of an ecclesiastical nature, but histories and biographies were written which were of a somewhat secular tone as well.

The Dark Ages usually refers to the period of time immediately following the fall of the Western Roman Empire in the year 476 A.D. Rome had been in the declining phases of its so-called glorious period for at least two centuries before its collapse. In 410 A.D. the Visigoths under a powerful leader named Alaric sacked the capitol city. Rome eventually recovered after Alaric's sudden death, but within the century had weakened and collapsed under its own impotent internal policies. The years following the fall of Rome brought a new group of people to the forefront of the literary scene in Europe. These people were of a Germanic lineage and had only recently adopted the Christian faith as their own. Christianity at this time was still a fairly new religion. It had gained a foothold during the reign of the Roman Emperor Constantine and had spread far and wide, eventually engulfing the pagan barbarians who had for years threatened Rome. In the centuries following Constantine's death, men of a divine nature like St. Augustine of Hippo became the most influential writers of the day. In 391 A.D. the Roman Emperor Theodosius forbade the use of any literary work that was not Christian. All other works were ordered to be destroyed, and the great library at Alexandria was burned to the ground. This was a crying shame, for it is a tragedy how many ancient works were lost in this fire. Perhaps it was this kind of behavior which led future historians to refer to this period of time as the Dark Ages.

One of the earliest of this new breed of Christian writer was the 6th century Bishop, Gregory of Tours who wrote a book called *The History of the Franks*. Gregory was one of those people who was fond of gossip. This was especially true when it came to unchaste women or hypocritical monks. However, he was also an astute and avid chronicler of signs and wonders, particularly of those things seen in the sky. One of the strangest fireballs that he mentioned occurred over Touraine France in the year 580 A.D. He said that it occurred in the 5th year of the reign of King Childebert. It was in the morning, just before sunrise, when a bright light shot across the sky disappearing in the east. This was followed by a loud sound as if trees came crashing to the ground. Gregory further added that this sound could not have been trees, for it had been heard for fifty miles around! From this report we can gather that this was almost certainly an exploding bolide.

Gregory mentions another puzzling prodigy that occurred during the year 580 A.D. He records that the area around Bordeaux in France was burned by fire sent from heaven. The fire, he said, was so fast that "homesteads and threshing floors with the grain still spread out on them were reduced to ashes." He reasoned that there was no obvious cause of the fire so it must have been a message sent by god.

On January 31, 583 Gregory made note of another fireball. He said that it was seen on a Sunday just after the bell had rung for Matins. A ball of fire fell from the sky and moved a great distance through the air. It was so bright that witnesses say it might have been high noon. When the fireball disappeared behind a cloud, the sky became dark again. Gregory also added that at the time this fireball was seen, the sky was overcast and it was raining. This event must have had plenty of witnesses, since, in those days people would walk to church and would have been much more observant of anomalies seen in the sky.

Another interesting report coming from the pen of Gregory occurred in the year 584 A.D. During the month of December when "the misshapen grapes appeared and the trees blossomed out of season . . . a great beacon was seen traversing the heavens, lighting up the land far and wide sometime before sunrise." He adds to this that "rays of light shone in the sky and in the northern sky a column of fire was seen to hang from on high for a period of two hours, with an immense star perched on top of it." Examining this closely it is likely that the "column of fire" was the northern lights, but the "immense star" leaves us wondering. Was Gregory referring to the planet Venus? Or perhaps Jupiter? Maybe, but we must remember that Gregory was merely the record keeper and not the actual eyewitness to many of the prodigies he recorded in his book.

Gregory mentions one more fireball which he says was seen either in 589 or 590 A.D. He says that "on a number of occasions fiery globes were seen traversing the sky at night which seemed to light up the whole earth." This might have been an early report of a meteor shower. Just before mentioning this he states that "In the same year so bright a light illumined a wide spread of lands in the middle of the night that you would have thought that it was high noon." It is not clear whether these two events are related, for Gregory often seems to dive right into another topic without first giving us closure of the subject at hand. There is one thing that interests me about this report. In a later chapter of this book I will discuss the strange nocturnal glows seen over Western Europe in early July 1908, and the glows witnessed by many people around the world after the volcanic eruption of Krakatoa in 1883. The cause of these glows was due to light reflecting off of dust particles in the atmosphere. This description of "light illumined a wide spread of lands in the middle of the night" is strikingly similar to reports from people immediately following the Tunguska blast of 1908, and the Krakatoa explosion a quarter of a century earlier. It is therefore possible that the "fiery globes" recorded by Gregory could have been the break up of a large meteor or comet. After impact with the earth a massive cloud of dust would have entered the atmosphere, and the winds would have carried it across the globe which could then have possibly produced the "light illumined" that Gregory tells us about. Of course, this light could also have been caused by a volcanic eruption somewhere on the planet, but at this time there is no evidence to back this up.

It is well known that meteor impacts or volcanic eruptions have in the past, and will in the future, cause mild climate disruptions around the earth. The Krakatoa explosion of 1883 was directly responsible for a median drop of 1.2 degrees Celsius around the globe. As a side note, it is recorded that in 589 A.D. major rainfalls and flooding occurred throughout much of Southern Europe. Whether or not this flooding was the direct result of a fireballs impact or not is not known, but it remains a possibility and is interesting to ponder.

Leaving Gregory of Tours we now turn to other major chroniclers who made their mark during the centuries which followed the fall of Rome. One of these was an obscure English monk named Bede who lived in the late 7th and early 8th centuries A.D. Bede's most famous work *An Ecclesiastical History of the English People* was finished in 731 A.D. It traces England's history from the time of Julius Caesar up until the age in which he lived. Bede's work is filled with fascinating accounts of early monarchs and leaders of the church in England. This includes a detailed narrative of the life of Augustine of Canterbury who is sometimes regarded as the father of Christianity in England, just as St. Patrick is in Ireland.

Like Gregory of Tours, Bede was fond of recording miracles and other portends that supported the tenets of the Christian faith. One such account takes place in the year 675 A.D. in the town of Barking. This small hamlet at this time in history boasted a cloister which housed a number of nuns. A plague had spread into the men's monastery but had not yet claimed any victims in the nunnery. The head sister was a pragmatist, and knew from experience that it was only a matter of time before the deadly disease took its toll on them as well. She therefore prepared her fellow sisters by asking them where they wanted to be buried when they passed on. The nuns were ambiguous, and could not decide where they wanted their final resting places to be. Bede then goes on to say that heaven decided to solve their dilemma for them. He states that "a light from heaven like a great sheet suddenly appeared and shone over them all." This light caused a degree of anxiety among the sisters, for they were unsure of its meaning. The nuns and monks both watched this light which seemed to linger for a minute before descending toward the south side of the convent where it seemed to pause for yet another short spell before doing something which defies explanation. It shot upwards and out of sight. There was no longer any question in regard to where the nuns were to be buried. The holy order of Barking would bury their deceased on the south side of the convent just as the light had signaled them to do. They were also somewhat more at ease to learn that their souls would be heading skyward, just as the light had.

This is one fireball which is hard to interpret. It was definitely not some common bolide breaking up in the rough turbulence of our atmosphere. Meteors do not linger, or pause, or change directions, and neither do they defy the Earth's gravity by shooting skyward. A meteor could explain part of this enigmatic fireball report. If the object seen by the nuns of Barking was a natural stellar body, it is possible that this was an early eyewitness account of what is called a "near miss." It could have been a large asteroid or comet that came a hairs length of slamming into the earth, but because of its tangent with the planet, and its phenomenal rate of speed, it somehow escaped the earths gravitational field. We can safely say that this object would have been seen by a great many people, and in all likelihood would have been extremely bright. This is probably why Bede described it as a "great sheet."

A good example of a "near miss" was the famous fireball seen on August 10, 1972 over much of the western United States and Canada. This object was identified as a good size asteroid which would have caused great devastation if it had actually struck the earth. This fireball was so bright that it was seen in the daytime, and was witnessed by hundreds, if not thousands of people. The Barking fireball could have been one of these monsters, but there is one thing about Bede's report that needs to be explained before we jump to any impetuous conclusions. First, the fireball seemed to linger, not just once, but at least twice. How can this be explained? The answer is it cannot, at least by any natural means. The law of inertia proves this. Simply put, it states that "A body remains at rest or moves in a straight line unless acted upon by some external force." Now, a meteor travels through space at a constant speed. However, when the rock enters the earth's atmosphere it would encounter turbulence which would slow it down. This would not be perceptible to the viewer on the ground because the change of speed would be minimal. The time it takes for the human brain to process this difference, in relation to the object seen by the eye would make it impossible to detect. I will give you an analogy that will show this. Imagine that you are standing on a freeway overpass and can see another overpass which is two miles away. You are able to see trucks passing over this overpass. A truck passing across there at 70 mph suddenly slows down to 40 mph halfway across. Would you be able to tell that the truck slowed down? Not unless you are Superman or the Six Million Dollar Man! However, the closer you get to that overpass the more likely you are to detect a change of speed. This analogy would seem to put Bede's report into question. If the Barking fireball was a meteor, either the eyewitnesses were mistaken in some of the details, or Bede screwed up the reporting job. It would have to have been an object of great size and have been real close for the report to be accurate. So if it was not a meteor, what was it? I remain perplexed. If the report is accurate, I can only conclude that the Barking fireball, whatever it was, was of intelligent design. I will say no more on the matter except to refer the matter to ufologists who might be interested in examining it further.

We now turn to one of the most famous fireball sightings of the Dark Ages. However, it became famous only because of who one of the eyewitnesses happened to be. This witness is synonymous with chivalry, and the effects of his reign are still being felt down to the present day. Charlemagne, king of the Franks was one of those rare characters in history where so much is written about him that he is often confused with being a mythological character. The real Charlemagne was a first rate soldier. During his reign he consolidated the Germanic tribes and put an end to the barbarian incursions into France and Italy which had been a constant thorn in the sides of many Popes. Pope Leo III crowned Charlemagne "Emperor of the West" in the year 800 A.D. for his services in the name of Jesus Christ. The irony of this is that no matter how strong and meaningful this great warrior was to the name of heaven, he was no match for a fireball from that very place!

Illustration 4

A contemporary biographer of Charlemagne named Einhard recorded in his *Vita Caroli* an event that took place in the year 810 A.D. Charlemagne was leading a campaign into Saxony against the Danish king Godefrid. Shortly before sunrise the army set out on the day's march. The king was riding along at the head of his army when the fireball struck. Einhard says that "Charlemagne saw a meteor flash down from the heavens and pass along the clear sky from right to left with a great blaze of light." This fireball must have been quite a sight for it spooked the king's horse which lowered its head and fell causing Charlemagne to topple to the ground. The force of the fall was so violent that it broke the buckle on his cloak, and his sword and belt were torn off. At the time he was thrown he had been holding a javelin, and this weapon was found nearly twenty feet away. Charlemagne was a stoic warrior, and this fall in no way shook his resolve. He continued with the mission as if nothing had happened. Perhaps a more timid ruler would have read some sort of omen into this mishap, but Charlemagne was no ordinary king. According to Einhard he refused to believe that this fireball had anything to do with the business at hand.

The period of time from Charlemagne in the 9th century until the Norman conquest of England two centuries later was a transitional period in European history. Most western countries had by this time been drawn into the web of the Christian faith. Those who were not were looked on with scorn and derision and no more than barbaric infidels who were destined to enter hell upon death. The medieval era had begun. This was the time of the castle and the moat. It was during this period in history that the Roman church's influence among the populace would attain it's highest level. It was a time of coming to terms with God. Nobleman and peasant alike lived their lives for the sole purpose of attaining salvation and the blessings of the Lord. This was also a good time to record fireballs which can be seen from the vast numbers of reports recorded by the steady hands of monks. These reports almost always had a foreboding sense of doom written into them, often associated with plagues, pestilence, and war.

One of the chroniclers of this period was the 12th century English monk Gervase of Canterbury. On November 29, 1177 he recorded a strange event which took place in the sky over Kent. It took place before the first hour on the vigil of St. Andrew. A red burning flame was seen in the sky which some people took to be that of "fiery dragons with many heads." The motif of the dragon fireball was a common one, for many people believed that this mythical beast really existed. Whether or not Gervase was one of these people is not known, for he may have been using the fire breathing dragon as a means of comparison, and may not have been using it in the literal sense at all. Whatever his reason was for using this metaphor, the meaning of the portent was clear. There was something foul in the air, and the god fearing people of Canterbury ought to prepare themselves for a period of trial and tribulation.

Another strange incident took place on June 23, 1178. Gervase says that "it occurred the day before the nativity of John the Baptist when the moon was full." He records: "from the east with the moon shining there sprung up a burning flame which threw forth sparks." He then adds to this that "the witness' were uneasy as the moon was struck hard and slayed." This is a somewhat ambiguous but interesting passage. Clearly there was observed a fireball of some sort which happened to appear in the same part of the sky as the moon. The question is . . . did this fireball really strike the moon? Of course, this is possible, but unlikely, due to a number of reasons. Firstly, the moon is 240,000 miles distant. So if someone were able to witness something striking it, that object would have to be gargantuan in size. The possibility that the fireball of June 1178 was one of these life-killer asteroids is remote, but not completely impossible. There is ample evidence of the moon being bombarded by meteors in its past. One only has to look through a small telescope and observe any of the large craters visible on its surface. The moon has no atmosphere, hence no erosion to wipe its surface clean. The craters on its surface are of an ancient age and have retained their integrity throughout the ages, quite unlike those on the earth which eventually succumb to erosion and plate tectonics.

Secondly, if the fireball recorded by Gervase had really hit the moon I believe that there would have been more written about it. Undoubtedly it would have created a massive cloud of dust which some astute monk or Chinese scholar would have recorded for posterity. No such record has been found, so I can only conclude that the likelihood of a collision on this date is remote. More than likely the fireball burned itself out at about the same time it happened to pass the moon, and gave the witnesses the impression that it "struck hard and slayed."

Illustration 5

Another chronicler of high repute during this era was Matthew Paris who left us a body of work which encompasses much of the 13th century. There are two reports that are worth mentioning here. The first one he mentions took place on the eve of the feast of St. James in the year 1239. Paris writes that "about dusk, before the stars had appeared there was seen in the blue sky a large star like a torch. This star rose from the south, and flew along a course which took it north. It was not a swift movement but graceful as a hawk flies." Paris then adds that "when it reached the middle of the firmament which is in our hemisphere it vanished leaving a trail of smoke and sparks in its wake."

Interpreting this report I must conclude this object to be a very bright bolide that happened to explode in the atmosphere. More than likely none of it , or at least very little of it reached the ground, for there is no record of anyone finding any debris as there was at L'Aigle, France or Weston, Connecticut many centuries later. Paris goes on to say that this star like object appeared like "a comet or a dragon which was greater to the eye than Lucifer . . . It took the shape of a mullet, very bright at the foremost part, but smoking and sparkling in the rear." This is the second time in this chapter where a fireball report is likened to a dragon. Throughout the course of researching this book I found many such references to the dragon. I therefore decided to devote a whole chapter on the subject, which I will include at the end of this book.

The fireball of 1239 as recorded by Paris almost certainly was a meteor. However, this cannot be said for a prodigy he tells us about which occurred some fifteen years later. This fireball or something like one was spotted on New Year's Day 1254 in front of a procession of monks at St. Albans who were commemorating the festival of St. Amphibalus. Paris says "at midnight, the air being most serene, and the sky covered with stars, the moon being eight days old, there appeared in the air marvelous to relate a kind of large ship, which was elegantly shaped, and equipped of a marvelous color." Paris goes on to say that "it stayed in the sky for a long time as if it were painted and in truth a ship made of planks, it slowly disappeared, and was believed to be a cloud . . . a marvelous and prodigious one."

This is one of the strangest reports that I have stumbled across in the course of my research. Was this a cloud as Paris believes it to have been, or was it something else? According to him, the moon was eight days old when this prodigy was seen. This means that it would have been at its waxing quarter stage. This would have illuminated the sky only slightly, but enough so that clouds could be discerned. However, Paris states that the sky was covered with stars, so this would seem to negate the cloud theory, or at least diminish it. Also, he describes it as having a "marvelous color." Could it have been the Aurora Borealis? However, usually this phenomenon is only seen in the higher latitudes. The fact that it stayed in the sky for a long time seems to throw doubt on the fireball theory, but due to the lack of evidence one way or another this possibility cannot be ruled out. Also, Paris is not exactly clear as to how long "a long time" happened to be. ufologist Jacques Vallee writing in the 1960's includes this report in his book *Anatomy of a Phenomenon: Ufo's in Space.* Could this have been an early sighting of a UFO? Nothing more is written on the matter, so the identity of the cloud-like object seen by the monks of St Alban remains to this day a mystery.

The late 13th and early 14th centuries were also the time when another somewhat enigmatic chronicler appeared on the scene. He went by the name of Matthew of Westminster. He is supposed to have been the author of a work entitled *Flores Historiarum.* However, it is now known that no one writer by that name ever existed. It is now believed that the work attributed to that man was really the efforts of many chroniclers who lived during those years. In other words, Matthew of Westminster is a pseudonym for a number of different unknown writers. Whoever wrote the *Flores Historiarum* dutifully recorded a number of encounters with fireballs. On July 29, 1263 it is written that "a certain very stupendous and wonderful sign appeared in the sky in the northern parts about midnight." Another fireball was reported seen over Uxbridge England in the early evening of November 4, 1322. The author states that:

> There appeared in the air to many beholders a wonderful sign.
> For a certain pile of fire the size and shape of a small boat, pallid,
> but of livid color, rising up from the south and crossing the

firmament with a slow and grave motion, set its course towards the north. Out of the front of this pile another very fervent fire of a red color and of greater quantity, similar in shape to the former, burst forth immediately with bright beams and great speed, flying through the air, which were seen quickly meeting against each other by many beholders and by turns frequently approaching with collisions and engaging in fearful combat, the blows of which conflict and the sounds of the crashes were heard at a distance from the beholders.

It can be plainly seen from this last sentence of this 700 year old fireball report that war was a familiar part of life in those days. Fireball reports during the Dark Ages and medieval times often equated these celestial bodies with "the clash of arms" or "fearful combat." For instance, a report in the *Chronicon Angliae* from 1360 uses a metaphor to describe what was either a meteor shower or the northern lights. It says:

In the summertime of this year, in flat and deserted places in England and France, and often visible to many, there suddenly appeared two towers, from which two armies went out, one of which was crowned with a warlike sign, and the other was clothed in a black color; they met and the soldiers overcame those in black, and a second time the warriors overcame the blacks, and returned to their tower, and the whole vanished.

Illustration 6

The 19th century British chronicler Edward J. Lowe described an event from the year 1366 which may in fact be the same report as the one described in the *Chronicon Angliae* in 1360. This report says that "There was a movement of the stars such as men never before saw or heard of; and those who saw it were filled with such great fear and dismay that they were astounded, imagining that they were struck dead, and that the end of the world had come." This report sounds like it could have been a meteor shower, perhaps a phenomenal one something in the line of the great Leonid shower of 1833 which reportedly contained numerous fireballs that rivaled the moon in size.

The last fireball that I care to mention in this chapter is probably the most famous one to hit the earth in medieval times. In the last chapter I made mention of a meteorite which has been kept at a Shinto shrine in Japan since its observed fall in 861 A.D. I have previously stated that the honor of the oldest observed fall in which evidence in the form of a meteorite exists belongs to the Nogata fall. However, this was not known to the western world until the late 19th century. Previously, the honor was held by the famous fireball which fell on Ensisheim in France on November 7, 1492.

Ensisheim is a small city in the Alsace province of France which has an ancient history, but is perhaps best known for what took place on that autumn morning in 1492. Between 11:00 A.M. and noontime a loud sound like thunder was heard above the city accompanied by a great ball of fire. The meteorite struck a wheat field near an area known as Gisgaud. Although there were a lot of witnesses only one person, a child, saw where the object struck the earth. The child led the local authorities into a field where much to their amazement they discovered a large crater five feet deep in which there was a stone which was estimated to weigh between 200 and 260 pounds. The stone was removed and said to have then been broken into many pieces by souvenir hunters. One of the largest chunks was given to the local church by the Hapsburg king Maximilian I who later became Holy Roman Emperor. Another piece was alleged to have gone to Duke Sigismund of Austria. Maximilian, it is said, gave the stone to the local church in the town where it was kept for many years. After the church adopted it, the stone was no longer subject to the sticky hands of would be treasure hunters. Today, the Ensisheim meteorite can still be seen in the Regency Palace at Ensisheim where there has been an annual meteorite exhibit held every year for the last decade. We now take our leave of fireballs from "The Age of Chivalry" and move into a period of time known as "The Age of Reason."

Chapter 5

FIREBALLS IN THE AGE OF REASON

The significance of the year 1492 is known even to those people who are not students of history. Ask the average person what happened during that year and they might come back at you with an old childhood nursery rhyme "In 1492 Columbus sailed the ocean blue." Even if they never heard of that old rhyme they surely have been made aware of a holiday in October which is named after the famous Genoese mariner. Christopher Columbus' reputation has suffered somewhat in recent years due to his involvement of the enslavement of some of the indigenous peoples of the Americas along with the diffusion of the notion of European cultural superiority which some historians have regarded as starting with him. This might be so, but what is known for sure is that the native population of the Americas were untouched by European influence until his arrival off of Hispaniola in October of 1492. Whether or not this was a good or bad thing I suppose rests with the individual; however, if it had not been Columbus then eventually it would have been someone else. The inevitability of a clash of continents and cultures had been in the makings since the first people crossed over the Bering land bridge thousands of years earlier. Isolationism is never a sure thing. It was only a matter of time before the oceans would be traversed.

I ended the last chapter with the fall of the Ensisheim stone. Ironically, this event occurred only one month after Columbus landed in San Salvador on October 12, 1492. I have therefore taken the liberty of using this crucial year in history and using it as the boundary line that separates "The Age of Chivalry" with the "Age of Reason." No matter what the reader may think of Columbus there is no doubt that his Atlantic crossing in 1492 ushered in the great age of exploration which sparked a new interest in the sciences and philosophy. It would only be a matter of time before the world view would change for good. Thirty years after Columbus' voyage, the Portuguese navigator Ferdinand Magellan, who was sailing for Spain, proved beyond a shadow of a doubt that a ship sailing west using the same latitude would eventually find itself back in its home port. It took Magellan three years, and he started with five ships in which only one returned with a skeleton crew minus its commander-in-chief, but the first circumnavigation of the world was completed. However, it was Columbus that got the ball rolling, and it was Columbus standing on the deck of the *Santa Maria* that observed the first recorded fireball of "The Age of Reason."

It happened on the evening of October 11, 1492, on a Thursday, at approximately 10:00 P.M. Columbus' ships had taken a southwesterly course two weeks earlier in the hope of running into "Cathay" (China) which he believed lay somewhere past the horizon. He had no way of knowing that a continent and another ocean lay between him and his destination. In fact, when he did eventually reach land the next day he was certain that he had succeeded in his endeavor. Columbus' small fleet of three ships had not seen land since leaving the Canary Islands back on September 06. For the men of this voyage, they were literally heading into the great unknown. One can only picture the scene some 500 years later. Columbus standing on the sterncastle of the *Santa Maria* peering out into the vast lonely darkness that lay ahead of them to the west. Was he correct? Would they reach Cathay or Cipango (Japan) by heading west instead of east, or had he duped himself into believing that such a thing was possible. He must have begun to wonder whether or not some of the superstitious members of his crew were correct. Would they sail into a murky sea where there lurked certain serpent-like monsters that could grab hold of a ship and carry it down into the watery depths? Or even more startling, would they merely sail off the edge of the world into a great chasm that lacked a bottom? The crew had almost mutinied three days earlier, but Columbus had managed to placate them. However, he was now at the end of his tether. If land was not spotted in the next couple of days, he would be forced to turn back, or worse . . . thrown into the sea by a disgruntled crew. It would be an ignominious end of a journey that had started with so much promise. Columbus was probably hashing these things over in his mind while he stood there stoic-like on the deck of his flagship. Perhaps he could see the outline of the *Nina* or the *Pinta* against the backdrop of an endless and monotonous horizon. Suddenly something caught his attention. His eyes captured a light low in the western sky described as "a little wax candle bobbing up and down." Excited, he called out to his loyal servant Pero Gutierrez who confirmed that the Admiral's vision was true. Columbus later wrote that this light was "moving up and down." Another man, Rodrigo Sanchez was summoned to bear witness, but apparently by the time that he arrived the light had vanished. Columbus was overjoyed and ordered a salvo of the guns. Four hours later at 2:00 A.M. a crewman on board the *Pinta* cried out that he could see land in the distance.

The mysterious light seen by Christopher Columbus has been the subject of debate for centuries. What did he see that night? It has been suggested that he might not have seen anything. The light was imaginary, and Columbus was only trying to revive the downcast spirits of his crew. Perhaps this was his way of feeding them a glimmer of hope. However, this does not explain the second witness, Pero Gutierrez, who also saw the light.

Washington Irving was best known for *The Sketchbook*. It contained morbid and sometimes fantastical tales such as *Rip Van Winkle* and *The Legend of Sleepy Hollow*. However, he was also a biographer of Columbus. He was convinced that the light Columbus saw that night was attributed to humans. He wrote:

> Columbus called Rodrigo Sanchez of Segovia and made the same
> inquiry. By the time the latter had ascended the round-house,
> the light had disappeared. They saw it once or twice afterwards
> in sudden and passing gleams; as if it were a torch in the bark

of a fisherman, rising and sinking with the waves; or in the hand of some person on shore, borne up and down as he walked from house to house.

Another 19th century biographer Justin Winsor was a bit more skeptical. He added a little arithmetic into the equation, which, if anything, showed that the mysterious light could not have come from land. Winsor believed that Columbus had to have been at least 12 to 14 leagues from the island that he would inevitably land the next morning. Using the lower of these numbers this would mean that the *Santa Maria,* with Columbus perched on the sterncastle, was at least 36 miles from San Salvador when he supposedly spotted the light. Given that distance and the low elevation of San Salvador it appears highly unlikely that the light he saw came from that island. In fact, due to the curvature of the earth it would seem almost impossible. The possibility arises however that the light was caused by a fisherman in a canoe, but the canoe would have had to have been a good ways from the shore. Also, at 10:00 P.M. it would seem unlikely that any fisherman would be that far from the shore, but the possibility cannot be totally ruled out.

So if we can almost positively say that the light was not the result of some human traveler, we need to turn our attention to natural phenomena. It is possible that the light could have been a star or a planet glimmering on the horizon, perhaps distorted somewhat by atmospheric conditions. However, we cannot forget that Columbus was a professional mariner. He was well acquainted with the night sky, and if the light had been a star or planet, he surely would have recognized it for what it was. At this point in the journey the three ships had reached the 24th parallel. The sky would have changed somewhat from the patterns familiar to the higher latitudes of Europe. However, Columbus was no stranger to this latitude. The Canary Islands lie at roughly the 25th parallel, and he was familiar with the sky around those islands. Due to this reasoning it seems likely that the light he saw was not a star or a planet. This leaves only a few more possibilities. It could have been ball lightning or a meteor. There is no evidence of any lightning storms that evening. Surely Columbus would have mentioned that in his journal. Since he does not mention it, we can safely rule out Ball Lightning which is almost always related to stormy conditions. We can therefore surmise that the light Columbus saw that evening could very well have been caused by a meteor. There are a few aspects of this sighting which support this theory. First of all, and most obvious, is the short duration of the light. Columbus only had time to call one witness to his side to confirm the sighting. By the time that Rodrigo Sanchez appeared on the scene the light was extinguished, or at the least, no longer visible to the men on the ship. Secondly, the light was described as "moving up and down." or "like a little wax candle." Fireballs are sometimes said to flicker, especially one with long trains that give off sparks. At first glance it would seem that the description of "a light moving up and down" would negate the fireball theory. This would be true if the fireball was witnessed from the land, but we must remember that Columbus was standing on the deck of a ship. This perceived movement can easily be attributed to the ships rolling on the waves which would naturally change the elevation of an object seen close to the horizon. In actuality the light was probably a lot higher than Columbus and Gutierrez perceived it to be since they were probably quite a distance from it.

The mystery of this strange light might have been solved if Columbus had thought to ask the natives if they had seen any strange lights in the sky the evening before. If he did, it is nowhere recorded, and most likely he did not bother to ask, for clearly Columbus believed that the light was caused by some torch or fire on the island. Therefore, the possibility of a fireball from the heavens probably never entered his mind. Unless new evidence arises, perhaps in the form of a contemporary journal from the time hidden away in some dusty corner of a moth-eaten library, it appears that we can only speculate on the matter. The true nature of Columbus' mystery light will probably never be known for sure, but a fireball makes as much sense as anything and is the most logical conclusion that I can come up with.

Illustration 7

We now take our leave of the year 1492 and turn our attention to the next few centuries that followed the landing of Columbus in the so-called new world. This was the period of exploration. The nations of Europe led by Spain, France, Holland, Portugal and England took to the high seas and began a systematic colonization of the continents of Africa, North America, and South America. The competitive nature of this struggle to gain the upper hand in trade led to the unfortunate conquests of many native peoples who just happened to get in the way of the players. The avaricious behavior of these trade giants even led Pope Alexander VI to divide the uncharted world in half giving half to Spain and the other half to Portugal. As if the other countries of the world would humbly sit back and accept the Pontiff's absurd diplomatic maneuver!

The 16th century was rife for a schism in the church, no thanks to Pope Alexander's ridiculous treaty. Not surprisingly, it was England, Holland and what is now Germany that led the various factions which separated themselves from the Church of Rome. This was accomplished through dissidents like Martin Luther and John Calvin who because of their outspoken opposition to the church were branded as heretics and excommunicated.

In science, the age old geocentric view of the universe finally succumbed to the Copernican heliocentric system. The telescope perfected by Galileo made it possible for man to see just how small he was in relation to the universe. During these years, however, one thing did not change. People were still seeing fireballs, and not surprisingly, reports were still similar to ones that had been reported by their ancient predecessors. The times change, but fireballs remain the same. They are a known quantity, a constant, much like rain. It is hard to predict when it might rain next, but rain it eventually will. The same is true of the fireball. It will only be a matter of time before one graces us with its presence.

Near the river Abdua in China in 1516, a remarkable event is supposed to have occurred. It is said that 1200 stones fell from the sky which somehow was calculated to have an accumulated weight of 120 pounds. At first glance this fireball report reeks of suspicion, for it seems incredible that all of the stones could have been collected and weighed with any type of accuracy. However, it may be the case that this was only an estimate of the true weight. It is said that one of these stones weighed 27 pounds.

On April 28, 1540 a large stone the size of a barrel fell at Limousin in France. This stone was said to have caused a crater several feet deep. It seems that the French had a rash of fireball sightings during the 15th, 16th and 17th centuries. This is no big surprise as France has seemingly always taken a leading role in the advancement of the sciences, and it is likely that they were a bit more conscientious at reporting these aerolites. The Provence area of France seems to have been particularly hospitable to fireballs in the early years of the 17th century. A bright one careened across the sky there on November 27, 1627. The recovered stone is said to have weighed 59 pounds. Ironically a similar event occurred almost exactly ten years to the day when on November 29, 1637 at 10:00 A.M. witnesses observed a fireball fall somewhere on Mount Vaison in Provence. Sources vary as to the exact weight of this stone upon recovery, but it can safely be said to have weighed between 38 and 54 pounds. It was said to have been of a black metallic color and was about the shape and size of a human head.

During the era of the renaissance it appears that fireballs must have taken sides with the reformation, for there are two instances of Catholic monks being killed by them. The first tragedy took place in September 1511. A large fireball was seen passing over the sky of northern Italy. The fireball burst in the air and scattered stones over the city of Crema in Lombardy. A monk was killed by one of these stones. It is said that ten of these stones weighed at least 100 pounds apiece, and after a thorough search there were found to be over 1200 of them that had been gathered up.

The second unfortunate monk was also killed in Italy, this time in Milan, when on March 30, 1650 a Franciscan priest was struck and killed by a stone. However unfortunate Catholic men of the cloth were when dealing with these stones from heaven, they were not the only ones to fall victim to their capricious wrath. In the year 1647 at Rochefort in France a fireball crashed into a cottage causing the roof to collapse and kill two men who were in the house.

The New England theologian and historian Increase Mather mentions a frightening incident in his book *Remarkable Providences* which came close to killing a ship's Captain. On July 17, 1677 a sailing vessel Captained by Thomas Berry was headed to the island of Madeira. The ship was off the coast of Cape Cod in Massachusetts when all hands were ordered on deck to assist in securing the longboats and take in the sails due to a storm that was seen ominously lurking on the horizon. However, before the storm reached the ship, a great noise was heard which Mather describes as "not sounding like a single cannon shot, but as if great armies of men had been firing one against the other." At this time Captain Berry also claims to have seen "A very black object fly before him about the bigness of a small mast." After this it was total chaos. Captain Berry was rendered unconscious by the blast which shook the vessel. He was initially thought to have been struck dead. However, he soon recovered his senses and after a few minutes was able to assess the damage that the mysterious black object had caused. The main mast had been split in half and the deck and holds of the ship had been punctured by tiny holes. Apparently a few fragments from this object had struck one of the ships pumps, for it was soon found to be inoperable. Below deck the smell of sulfur was so strong that for thirty minutes no man could enter or he would be overcome by the fumes. Captain Berry realized that the most expedient thing to do was turn the ship around and head back to Boston where it could be refitted.

This incident is one that can be scrutinized up and down, left and right, and one would still come away feeling a sense of confusion. Probably the first thing that comes to mind when reading this account was that the ship's main mast was struck by lightning. After all, a storm was close by, and the ship's hold reeked of sulfur for some time after the blast. It is possible that the strike could have ignited some gunpowder stored in the hold. It also could have been the strong odor of ozone that one can smell before a storm hits. Perhaps the amount of time was merely exaggerated. This does not, however, explain the dark mast-like object seen by Captain Berry shortly before he was rendered unconscious. Also, what about the holes that penetrated the ship's deck and took out one of the pumps? This evidence clearly shows that the object was an exploding fireball. However, if it was, where was the evidence? There should have been stone fragments left behind, but Mather does not mention anything about stones being found in his report. I surmise that the "very black object" that Captain Berry saw was the smoking remnants of a large piece of rock. This object may have struck the main mast causing most of the damage. Since no rock was found by the crew, this object must have been carried by its momentum into the sea. The small holes were probably caused by smaller debris after the fireball exploded. Its effect would have been similar to canister or grapeshot being fired out of a cannon. It is a small wonder no one was killed.

The smoky, black death of a fireball is more common than one would think. The most famous example would have to be the Sikhote-Alin fall over Siberia in 1947 which I will discuss in a later chapter of this book. There have, however, been numerous other examples of smoky black objects falling from the sky. One of these occurred in July, 1766 near the village of Albereto in Northern Italy. The sky was said to be clear when at about 5:00 o'clock in the evening the normal peace of the town was interrupted by a startling event. The *Edinburgh Encyclopedia* gives a good description of what happened:

> About 5 o'clock in the evening, when the peasants were dispersed over the fields, engaged in their rural labours, there was suddenly heard, not only in Albereto, but in other places at a considerable distance to the west, and even at Modena, an unusual noise, like the discharge of artillery, succeeded by a whizzing in the air, like that produced by a cannon bullet when powerfully propelled. The Duke of Modena's gardener even believed that a cannonball was descending into the garden. Others either did not hear the whizzing noise, or had not paid attention to it. In Albereto, however, it was not only heard, but a body was moreover seen traversing the air with great velocity, and falling abruptly to the earth. To some of the distant bystanders it appeared in a state of ignition, but to two

ladies, who were within a mile of the spot, it seemed opaque and smoking. They instinctively clung to a branch of a tree, but an ox which was near them, fell to the ground from terror. The stone, which diffused an odor of sulfur, had penetrated the soil to nearly the depth of a fathom, was still hot when taken up, and had the appearance of a sandstone of great weight, of an irregular triangular figure, with its external surface uniformly burnished over with black, as if from the effect of fire.

This fireball report and the one related by Increase Mather nearly a century earlier bear similar characteristics. They both told of the strong odor of sulfur which accompanied the fireball. This seems to be common when a meteorite is examined up close and personal minutes after it strikes the earth. Both accounts also described the objects as being "very black" or "opaque and smoking." These were descriptions given by witnesses who were extremely close to the point of impact. The two ladies were close enough to see the object after the flames had been extinguished from its body. Captain Berry was probably a good deal closer, and was actually able to describe the size of the object which was probably a very accurate assessment.

The last two fireball sightings that I will relate in this chapter are probably two of the most famous falls of the 19th century. However they are not only famous, but also important to the study of fireball history. It was the fireball reports at L'Aigle, France in 1803 and Weston, Connecticut in 1807 that proved to the skeptics once and for all that rocks do fall from space. If there had been any doubt before these two fireballs, all this cynicism was wiped away by these two falls. As I have already mentioned, rocks have been known to fall from the sky since the early days of recorded history. However, contrary to all the evidence, at the turn of the 19th century there were still skeptics who balked at this notion. To the skeptics' credit, however, it must be noted that at that time very little was known of the Earth's atmosphere. It has only been within the last few centuries that the sky above us has been studied with any kind of scientific instruments. This was accomplished in a large part by sending up balloons which carried a payload that could gather this information. A major factor in this study was due in part to the French balloonists of the late 18th century who performed various experiments while at the same time attempting to set personal achievements which would leave them immortalized in history. One of these early pioneers was the French chemist, Pilatre de Rozier, who attempted to cross the English Channel in a balloon in 1785. His attempt ended in tragedy. As crowds watched in horror from the shores off Dover England the balloon he piloted disintegrated in a fiery explosion which sent the craft into a helpless plummet into the sea.

Early aeronautical researchers like Pilatre de Rozier ushered in a new age of enlightened thinking. One of these late 18th century pioneers was the German astronomer, Ernst Chaldni, who was criticized for espousing the theory of space rocks among the scientific circles that he traveled in. After L'Aigle and Weston, he would be laughed at no more. No longer could meteors be attributed solely to strong winds which propelled them skyward as Aristotle assumed two thousand years earlier. Other notions were also shelved. The fantastical concept of a living sea which orbited the earth in the upper atmosphere, and occasionally dumped rocks, fish and frogs from its supposed depths also found itself relegated to the bone yard of discredited theories.

I briefly mentioned the L'Aigle fireball back in Chapter 2. As I stated earlier this fall was the result of a large meteor which broke up in the atmosphere and showered this small farming community with thousands of stones. After an exhausting search, over 3000 fragments of this meteorite were found. The event took place on April 26, 1803 at approximately one o'clock in the afternoon. A letter from a Mr. Marais of L'Aigle to a friend living in Paris gives an excellent detailed account of what happened that afternoon:

> An astonishing miracle has just occurred in our district. Here it is without alteration, addition, or diminution. It is certain that it is the truth itself. On Friday last, 6th Floreal, between one and two o'clock in the afternoon, we were roused by a murmuring noise like thunder. On going out we were surprised to see the

sky pretty clear, with the exception of some small clouds. We took it for the noise of a carriage, or of fire in the neighborhood. We were then in the meadow, to examine whence the noise proceeded, when we observed all the inhabitants of Pont de Pierre at their windows and in gardens inquiring concerning a cloud, which passed in the direction of from south to north, and from which the noise issued, although that cloud presented nothing extraordinary in its appearance. But great was our astonishment when we learned, that many and large stones had fallen from it, some of them weighing ten, eleven, and even seventeen pounds, in the space comprised between the house of the Buat family (half a league north-east of L'Aigle) and Glos, passing by St. Nicholas, St Pierre, & etc, which struck us at first as a fable, but which afterwards was found to be true.

This account of the L'Aigle fireball is important in a descriptive way for its geographical references. However, Mr. Marais himself did not see the fireball which accompanied the noise he heard, although he did mention some clouds that he had seen in the sky. This was not the case with some other reports that came in. A man named Buat told people that a fireball was seen hovering over a meadow, but believed that it might have been a wildfire. A Mr. Lamarck stated that he had received some letters from eyewitnesses who recalled seeing a fireball passing across the sky from east to west with great velocity. The first person with any kind of scientific training to arrive on the scene was a young 29 year old physicist and astronomer named Jean Baptiste Biot. Biot was a former revolutionary and an avid supporter of Napoleon Bonaparte. He was also one of those pioneering French balloonists, and would later set an altitude record when he and a colleague ascended to a height of 13,000. The purpose of their flight was to study the magnetic, electrical and chemical properties of the atmosphere at higher altitudes.

After hearing about a possible rock fall from the sky at L'Aigle, Biot left Paris on a mission. He was to gather reports and samples and try to determine all of the facts related to the incident. His first stop was at the town of Alencon which lies about 40 miles southwest of L'Aigle. Here he gathered eyewitness reports, and from these was able to determine the fireball's general direction of travel. In this instance most of the eyewitnesses agreed that the fireball was seen moving in a northerly direction followed by a great noise. Biot left Alencon and accumulated more reports from the various towns and hamlets that dotted the Norman countryside around L'Aigle. Most of the people he talked with had a similar story to tell in that there was a fireball seen after which a sound similar to a cannon was heard for a period of five to six minutes, followed by a rumbling noise. This noise was heard in a 90 mile radius of L'Aigle. The closer the witnesses were to the fall site the more they reported the fireball was accompanied by a "hissing sound." Biot gathered some of the stones which he brought back to Paris for study, including one which weighed 17 pounds.

There were numerous stories of the fireball floating around the vicinity of L'Aigle. Indeed, it was the talk of the town. Some of these reports were fantastical, and were probably slightly embellished for effect. It was the hot topic at the local inns and taverns, places where people could gather around a fire and see who could tell the biggest tall-tale. One of these reports came from the village of La Sapee and was in all likelihood probably first told at the local tavern by a wag to a table of inebriated patrons. It was probably at the expense of the local dolt who may or may not have been present to laugh at his own mockery. He is alleged to have said, "Good God! Is it possible that thou canst make me perish thus? Pardon, I beseech thee, all the faults that I have committed." Although the witnesses to this supposed statement might have laughed at it after the danger had passed, they themselves were probably thinking something along a similar line when the danger was still present.

In another report, a priest stated that he had seen one of the stones fall and that it was accompanied by a hissing sound. The stone is said to have fallen next to where his niece was standing and he asked her to pick it up so that he could examine it, but she was afraid to touch it. This was not an uncommon reaction, for most people had never seen stones fall from the sky. However, they were well familiar with the biblical tales of locust plagues and falling frogs, so they were probably rather sheepishly reluctant to test God's motive from this ominous sign from heaven. Indeed, some people feared that the end of the world was at hand.

Although no one was killed by one of these stone fragments there is an instance of a man being struck by one. In the town of Avnees a wire manufacturer named Piche was outside working with some of his men when he was allegedly struck by a piece of the meteorite on his arm. It then landed at his feet and when he attempted to pick up the glowing hot rock he immediately dropped it. This report seems to be one that may have been embellished somewhat, but it cannot be completely dismissed out of hand.

These eyewitness reports along with many others were collected along with some of the stones which eventually found their way to laboratory tables. The stones were given careful examination and determined to contain high proportions of silica, magnesium and iron along with smaller quantities of nickel and sulfur. These were elements that were not well known in the L'Aigle landscape. Due to the overwhelming evidence that had been accrued at L'Aigle, even the most stubborn cynic could no longer deny that stones on occasion fell from outer space. However, there were still some people that needed to be convinced. North America lay an ocean removed from Europe, and in the early 19th century news traveled very slowly. Even when the news from L'Aigle reached America, the reception it received was condescending at best. It would take four more years and another fireball to convince these hard headed Yankees that stones fell from the sky.

It was on one of those cold New England mornings on the 14th day of December 1807 when America's most famous fireball crashed its way into history. At approximately 6:30 A.M. Merwin Burr was standing in the road in front of his house in Huntington, Connecticut when he heard a loud noise. At first he did not know what to make of it because it was still dark outside. However, a half hour later the sun had risen, so he could make a search for whatever it was that caused it. To his utter amazement he found that a stone had crashed into a rock only fifty feet from where he had been standing. The collision had shattered the stone into fragments.

The largest of these fragments he estimated weighed between 20 and 25 pounds. Burr picked this stone up and perceived it to be quite warm. He must have then glanced up at the sky with a sense of wonderment. He did not know it at the time but what he was holding in his hands was a piece of history. I say this in both a literal and a figurative sense. This stone had been traveling through space for millions of years, a fragment of a bigger mass that had broken up and scattered other fragments over a large area of Weston, Connecticut. Huntington, where Burr lived is located just north of Weston. Today, this event is known as the Weston meteorite.

The Weston fireball was seen over a wide area of New England, and observed as far south as New York. Reports were numerous, but most had a similar tale to tell. One of these reports came from a judge in Weston named Nathan Wheeler who happened to be passing through an enclosure adjoining his house when he observed the fireball. He was later interviewed by Benjamin Silliman who was a distinguished professor of natural history at Yale University. Silliman included Judge Wheeler's testimony in his report on the Weston fireball first published in the *Connecticut Herald* and later published in the *Boston Review*. The judge reported that he was drawn to a sudden flash in the sky which seemed to illuminate every object around him. The report states:

> Looking up the judge saw a globe of fire passing behind a thin cloud. This did not wholly hide the fireball from view as it was extremely bright, and was like seeing the sun through a mist. The fireball rose from the north and proceeded in a direction nearly perpendicular to the horizon, but inclining, by a very small angle, to the west, and deviating a little from the plane of a great circle, but in pretty large curves, sometimes on one side of the plane, and sometimes on the other, but never making an angle with it of more than 4 or 5 degrees. Its apparent diameter was about one-half or two-thirds the apparent diameter of the full moon. Its progress was not so rapid as that of common meteors and shooting stars. When it passed behind the thinner clouds, it appeared brighter than before; and, when it passed the spots of clear sky, it flashed with a vivid light yet

not so intense as the lightning in a thunderstorm, but rather like what is commonly called heat lightning. Where it was not too much obscured by thick clouds, a waving conical train of paler light was seen to attend it, in length about 10 or 12 diameters of the body. In the clear sky a brisk scintillation was observed about the body of the meteor, like that of a burning firebrand carried against the wind. It disappeared about 15 degrees short of the zenith, and about the same number of degrees west of the meridian. It did not vanish instantaneously, but grew, pretty rapidly, fainter, and fainter, as a red hot cannon ball would do, if cooling in the dark, only with much more rapidity. There was no peculiar smell in the atmosphere, nor were any luminous masses seen to separate from the body. The whole period between its first appearance and total extinction, was estimated at about 30 seconds. About 30 or 40 seconds after this, three loud and distinct reports, like those of a four pounder, near at hand were heard. They succeeded each other with as much rapidity, as was consistent with distinctness, and, all together, did not occupy three seconds. Then followed a rapid succession of reports less loud, and running into each other, so as to produce a continued rumbling, like that of a cannon ball rolling over a floor, sometimes louder, and at other times fainter: some compared it to the noise of a wagon, running rapidly down a long and stony hill; or, to a volley of musketry, protracted into what is called, in military language, a running fire. This noise continued about as long as the body was in rising, and died away apparently in the direction from which the meteor came.

Judge Wheeler had a great view of the fireball as did his neighbor Elihu Staples who observed, "A streak of light passing over his orchard accompanied by a noise which sounded like a whirlwind, and three loud explosions which occurred with each leap of the fireball." The residents of Weston had never seen anything like this before. News spread quickly, and after the first stones were recovered, a carnival like atmosphere ensued. It became a scavenger hunt of sorts as people flocked to the fields and woods where the stones were thought to have fallen.

William Prince was still in his bed when the fireball struck. He heard loud explosions followed by a noise which he thought sounded like a heavy object hitting the ground. As it was still dark he was not able to investigate the matter right away. In fact, he thought nothing more about it until he went into town later that day and found everyone buzzing about it. He immediately returned home and after a thorough examination of his land he found a hole not more than 25 feet from his house which contained a stone estimated to weigh around 35 pounds. Unfortunately for posterity, Prince was a sort of opportunist and broke the stone up selling fragments of it to souvenir hunters. In the end, all that was left was a chunk weighing 12 pounds which subsequently disappeared.

Only 2 miles south-east of William Prince's residence lived Ephraim Porter and his family. Like Prince, they too heard the fireball but did not actually see it. They did, however, like most of their neighbors, investigate the noise and were perplexed to see smoke rising from a certain spot not too far from their house. Porter concluded that the smoke was probably the result of a lightning strike and forgot about it until he heard that people had found some stones that had fallen from the sky. He then investigated the place where he had seen the smoke and much to his amazement found a hole in the ground about 2 feet deep and 20 inches in diameter. Interestingly, the perimeter of the hole was covered with a fine blue powder. It is not known what caused this, but one can surmise that it might have been residue from the heated stone that was sheered off when it struck the earth. Some metals are known to give off a bluish or green smoke when heated to high temperatures. It is also common for some metallic ores to appear blue or green after undergoing smelting procedures. This process can readily be seen when examining the slag left over after being dumped by a crucible in foundry applications. Porter was able to extract a number of small rock fragments from this hole which all together weighed about 20 pounds.

Professor Silliman believed that the three loud explosions heard by witnesses were directly related to places where the stones were known to have fallen. Therefore, each explosion represented a general area of fall with a wide expanse of land in between each. The first leap of the fireball mentioned by Elihu Staples was therefore responsible for the stones which fell in the north part of Weston and the town of Huntington around the home of Merwin Burr. Progressively, the fireball moved south until experiencing a second explosion centered around the vicinity of William Prince's residence. The third and final leap of the fireball was its death song. Silliman believed that these stones fell in a field owned by a local farmer named Elihu Seeley. Seeley found a number of fragments which had caused some of the earth in his field to be upturned as if a plow had recently gone through.

Thus far we have examined the accounts of witnesses who were residents of the town of Weston or its nearby communities. However, as I have already stated, this fireball was seen over a large area of the New England states and New York. One of the most distant witnesses was a woman by the name of Mrs. Gardner who lived in Wenham, Massachusetts, a town nearly 180 miles from Weston. She was an early riser, and part of her morning routine after getting out of bed was to go to the window in her bedroom and look out the window to observe the weather. Mrs. Gardner stated that it was about a half hour before sunrise, and was looking up at the sky when she observed a disk about the size of the full moon moving over her barn. The disk was running a course roughly parallel with the horizon and was visible for about thirty seconds.

Illustration 8

Another report came from a distance which was even further away. William Page, a lawyer, was standing in the doorway of his home in Rutland, Vermont when he observed the Weston fireball. He gave a detailed description of his experience to a Professor Hall of Middlebury College. This account was later given to the famous oceanographer Nathaniel Bowditch who is perhaps best known for his book *The New American Practical Navigator* which even after 200 years the U.S. Navy still carries on board its ships. In 1811 Bowditch published Page's account in *The Journal of Natural Philosophy, Chemistry and the Arts*. The report from Page is as follows:

> I was at the west door of my house on Monday morning the 14th of December, 1807, about daylight, and perceiving the sky suddenly illuminated, I raised my eyes, and beheld a meteor of a circular form in the southwesterly part of the heavens, rapidly descending to the south, leaving behind it a vivid, sparkling train of light, the atmosphere near the south part of the horizon was very hazy; but the passage of the meteor behind the clouds was visible until it descended below the mountains, about twenty miles south of this place.

The account of William Page is interesting if only because of the distance involved. It is 220 miles from Rutland to Weston. Correlating the Rutland and Wenham observations with those reports coming from the actual place of the fall, Bowditch was able to calculate a mean trajectory of the fireball. The work of Silliman and Bowditch can almost be said to have brought America into the early space age. Space became alive. There was now more to it than just a vast emptiness occasionally dotted by a distant star, planet, or passing comet. If rocks abounded in this mysterious region, what else may lurk out there?

The Weston fireball of 1807 marked the end of an era— one which started with the first reported sightings in the ancient world more than three thousand years earlier. After 1807 most fireballs could be readily identifiable as meteors, but there were others that could not be explained so easily. There were some fireballs that seemed to defy the laws of physics. In the last chapter I mentioned a fireball seen over the monastery in Barking, England which was recorded by Bede. This fireball was unique in the way it behaved in relation to the so called "weakest force" gravity. I have also mentioned several other fireballs that cannot possibly be attributed to a meteor. For instance the "great beacon" mentioned by Gregory of Tours which took place in the year 584 A.D., or the large ship related by Matthew Paris as having appeared in the heavens in 1254 A.D. So from these enigmatic reports we can see that strange things that appear in the sky do not always have to be meteors, although a substantial portion fall into this category. The farther back in time that we go the sketchier the reports get. This, of course is quite understandable. Language changes, and time has a way of erasing even the strongest memories. Some reports are even so laconic that it is frustratingly hard to tell what the prodigy might have been. A good example would be one mentioned by Julius Obsequens which was said to have occurred over Gaul in the year 113 B.C. The brief entry for that year states simply "The sky appeared to be on fire." I guess that we can thank Mr. Obsequens for leaving us this little tidbit of information, for at least we know that something was seen in the sky over Gaul in 113 B.C. However, we are at a loss to say just what. The sky on fire could pertain to a number of things. It certainly could have been a meteor, or it could have been the Aurora Borealis (however, at that latitude extremely unlikely), or an atmospheric phenomenon known as a sundog which more correctly termed is called parhelion, which is nothing more than reflected sunlight from ice crystals in the clouds. In the present day and age we are equipped with better science, and can utilize the tools that have been given to us by the persistence and genius of our predecessors. We will next examine fireballs in what I term "The Modern Era."

Chapter 6

FIREBALLS IN THE MODERN ERA

I ended the last chapter talking about the fireball which struck Weston, Connecticut in 1807. Therefore, I think that it is appropriate to begin this chapter at that point. Recently the citizens of Weston celebrated the 200th anniversary of that event. Since that time the observation of fireballs has become a past time of sorts. The dates of various meteor showers are well known, and sky watchers can plan their schedules accordingly in the hopes of seeing a spectacular show.

Since I have decided that the modern era of fireballs began 200 years ago, I will relate a few instances of spectacular ones that were witnessed in the 19th century. This was an important time in fireball history. The nature of these objects was now better understood. Chroniclers were joined by men of science who recorded these reports and made earnest attempts to find the evidence that these balls of fire left behind. There was no longer a question of whether or not rocks fell from the sky. The controversy was over. L'Aigle and Weston proved that. The debate now raged on just where in the solar system the rocks came from. One mode of thinking believed that they must form somewhere in our atmosphere. The February 1843 edition of *The American Eclectic* writes:

> In respect to fireballs which frequently appear in connection with
> the falling stars, are certainly nearly related to them, and allow
> us a better opportunity of investigation, as they are larger and

approach nearer, we consider the following facts established . . .
Fireballs are usually seen just when they have reached their
greatest brilliancy, and burst asunder. Their beginning has been
seldom observed. In such cases, however, which Chaldni has
carefully designated, there first appeared a distant streak of
light, which, as it approached nearer, rounded itself into a ball;
sometimes, also, several streaks, and these, too, crossing each
other. This seems to indicate that the balls, as such, are formed
in our atmosphere, and that the matter composing them flows
together into a ball, from different directions, out of the higher
regions of space, without having been previously united, or
pursued its path anywhere in the firmament, as an independent,
substantial body.

Of course, there were other types of fireballs besides meteors that
science was also attempting to gain a better understanding. The list includes
such phenomena as ball lightning, St. Elmo's fire and the will-o'-the wisp.
This led to some confusion, but at least by this time period they were able
to differentiate each as distinct and separate phenomena as the article
from *The American Eclectic* states:

A phenomenon observed in tempests, seems not unfrequently
to have been confounded with fireballs: namely, flames—such
especially as rise up out of the earth or sea, but also those
which, falling, downwards sometimes assume the form of
perfect fireballs, and move on, then in a moment stop again . . .
In 1822, in the vicinity of St. Omer, fireballs were thrown out
by a water spout. These electrical phenomena and gaseous
appearances on the surface of the earth and of the sea, must
not be confounded with those great phenomena, which come
down from unmeasured heights, and throw whole continents
into alarm.

As the 19th century dwindled away so too did the superstitions attached
to fireballs. After all, the knowledge of our universe was expanding.
The church had finally lost its hold on the universities, and the men of

science were now of a more secular proclivity than their predecessors. The mystery of the atom was unraveling, and the progenitors of quantum mechanics were building the foundation for a revolution of thought that came in the decades ahead. However, although some of the mystery linked with fireballs had been untangled, the excitement and star power of their performance had not diminished. If anything, the thrill of seeing one increased since the stigma of bad omens attached to them had subsided. With these irrational notions relegated to a bygone day, the observers could now bathe themselves with the beauty of these natural pyrotechnic displays, and rays of cosmic splendor without fearing the displeasure of a vengeful god, or worrying about a bad harvest or a coming plague. One such report made its way into *The Sheffield Times* and was recorded by William Lackland in his translation of the French work by Zurcher and Margolls entitled *Meteors, Aerolites, Storms, and Atmospheric Phenomena*. It took place in Hurworth, England in October 1854. A "gentleman" and his brother, whose names have seemingly been lost to history, were traveling home one evening when they witnessed a spectacular fireball. The report, as translated by Lackland is as follows:

My brother and I were returning home at nine o'clock in the evening, and had just reached the end of the village, and were about to cross a meadow of considerable breadth. The sky was clear and starry, but dark. We were looking at one of the brightest constellations, when, at the very point at which our gaze was fixed, we beheld a magnificent sight. A cry of admiration and astonishment escaped us both. What we saw was a globe of fire, at least double the size of the moon when it rises. It was as red as blood, and shot out sparkling rays, which were marked in sharp outlines, as old engravings represent the rays of the sun. It drew after it a long trail of light of the most beautiful limpid golden color. The train had no resemblance to the hairy tail of a comet, but was more like a solid column, of great breadth and perfect compactness, standing out against the deep blue of the sky. In the beginning it presented the appearance of a straight line, but as it moved the heavens, it described the curve of an arch with sparkling scintillations of great intensity which however did not pass beyond the well defined exterior line. Its direction was from northeast to south west and its length so enormous that when its nucleus was disappearing under the southwestern horizon, the trail was still visible at the northeast in all its original splendor . . . When this globe of fire was immediately above us, it seemed to pause for a moment with vibrations so violent that I was afraid it would fall on us. But, the next instant, I saw that the vibration was only a whirling motion, and that it was turning rapidly on its axis, passing from a vivid fiery red to the deep red mentioned above, without, however, losing anything of its general appearance. We continued to see it, looking as brilliant as ever, behind the trees on the other side of the village. While this globe was passing over us it seemed a little smaller than when it first appeared on the horizon, no doubt because of its great elevation, just as the sun and the moon look smaller at their meridian than when they are rising . . . As I have been, for a long time past, in the habit of watching the stars, I have seen several brilliant meteors, but never any that could bear the least comparison to this one, whether for dimensions or for splendor and duration. Owing to its height in the air, it must have been visible at a great distance, and I hoped that it would have been seen and described by intelligent observers. As such, however, has not been the case, I have thought it my duty to furnish some details concerning a phenomenon so grand and striking.

This was quite a remarkable and detailed account of a fireball, the likes of which, unfortunately, are uncommon even today. The writer not only gave the direction of travel and a size comparison of the object, but spiced the account up with descriptive Victorian prose. After all, how many people today would release a "cry of admiration and astonishment." The green fireball that I saw in January, 2007 left me astonished as well as an admirer of its "splendor", but I could not give a cry of admiration because I was rendered speechless!

Not all 19th century fireballs showed the zeal and colorful excitement of the Hurworth fireball. Some accounts were merely recorded for posterity in a laconic sentence or two. General Simon Griffin, a decorated veteran of the American Civil War, mentions in his *History of the Town of Keene* an account of a fireball seen over Keene, New Hampshire on the evening of March 2, 1875. He writes simply that "A brilliant meteor observed at 11:30 P.M. over the skies of Keene."

The 19th century was also host to one of the most brilliant meteor showers on record. On the evening of November 13, 1833 it is recorded that the sky was so full with meteors that "persons sleeping in rooms with uncurtained windows were aroused by their light." Witnesses say that they ranged in size from small points of light to fireballs larger than the full moon. One witness to this magnificent performance of the heavens was a plantation owner from South Carolina whose name I have not been able to ascertain. The following account was written in Daniel Kirkwood's *Meteoric Astronomy* published in 1873. It says:

> I was suddenly awakened by the most distressing cries that ever fell on my ears. Shrieks of horror and cries for mercy I could hear from most of the negroes of the three plantations, amounting in all to about 600 or 800. While earnestly listening for the cause I heard a faint voice near the door, calling my name. I arose, and, taking my sword, stood at the door. At this moment I heard the same voice still beseeching me to arise, and saying, O' my God the world is on fire!' I then opened the door, and it is difficult to say which excited me the most,-the awfulness of the scene, or the distressed cries of the negroes, upwards of a hundred lay prostrate on the ground, some speechless, and some with the bitterest cries, but with their hands raised, imploring God to save the world and them. The scene was truly awful; for never did rain fall much thicker than the meteors fell towards the earth; east, west, north, and south, it was the same.

The great meteor shower of 1833 we now know was none other than the annual Leonids which occurs every November. While orbiting the sun the Earth passes the radiant point of these meteors, which as the name suggests happens to be located in the constellation Leo. One source estimated that the 1833 Leonids generated up to 34,000 shooting stars per hour. This is quite an extraordinary amount when one reflects upon it. An October meteor shower in the year 1875 estimated that at least 50 shooting stars were seen in the interval of one minute. A little arithmetic tells us that there would have been over 3000 seen in this shower over the course of an hour. Looking at it in this respect, it is easy to see how the 1833 Leonids shower garnered so much attention.

It was around this time that astronomers figured out that meteor showers were an annual event. It was then assumed that these rocks were nothing more than the debris left behind by some passing comet in a bygone age. Since 1833 there have been several other infamous meteor showers including the 1966 Leonids which probably rivaled the 1833 shower in star power, but lacked the doomsday prophets of that older event. As the 19th century came to a close, our world had changed from one of candles and lanterns to the incandescent light bulb and the flashlight. The fireball reports, however, kept coming in. We had moved into the 20th century, which would be unlike any other in our history. If the 19th century was the age of the Industrial Revolution, the 20th century was the age of technology. This brought about a change in the way people viewed fireballs. In particular, it was the invention of the airplane that caused this change of perspective in the way we look at the sky. It is now almost impossible to look up into the heavens anywhere in the civilized world and not see an airplane flying somewhere within the dome of the sky. During the day we can make out the elongated grey forms, or the vapor trails left behind by these would be modern followers of Icarus. At night we can look up and see the ubiquitous blinking lights. There are also hundreds of satellites which traverse the region of space just outside our Earth's atmosphere. The space station and the space shuttle also add to the modern panorama of the sky. Indeed, over the last century the sky has changed drastically. Its natural look corrupted by the presence of man in flight. The only time in the last century in which the sky was devoid of airplanes was the few days following the 9-11 terrorist attacks in 2001. Even then, this was only confined to most of North America and parts of Europe. Ironically, this was a very quiet and peaceful time for our atmosphere notwithstanding the turbulent and chaotic times down below.

As horrible and meaningless as these terrorist attacks were, they did vault one important issue to the forefront of world news. This was the issue of airline safety. Airplanes can and do fall out of the sky, and on occasion sometimes slam into heavily populated areas of our ever shrinking planet. Whether it is due to a terrorist attack or the more likely culprit of some kind of mechanical failure, it is an inevitable fact, as somber as this may seem. It is because of this disturbing possibility that when a spectacular fireball is observed, one of the first things that comes to mind is that it might be an aircraft in distress. This is especially true when the fireball is accompanied by noise.

A good example of this occurred just recently on the evening of August 20, 2006. According to the BBC, a large fireball appearing something like a plane going down was spotted crashing into the sea off of the Hebrides. This fireball left a trail of smoke in its wake which caused many people to believe that they had witnessed some sort of aircraft going down. The authorities notified the Coast Guard, which then prepared for rescue and recovery efforts. Apparently, after consulting with aviation officials, it was determined that no aircraft was reported missing. The final word on the matter from a spokesman from Kinloss Airfield was that it was only a bright meteor associated with the Kappa-Cygnid meteor shower which peaks about the time the fireball was reported.

Crashing airplanes are not the only things associated with fireballs these days. The first satellites were launched into space fifty years ago. Since that time we have created a sort of space graveyard. There are hundreds of man-made objects currently in orbit around our Earth. Every so often some of this space junk uses gravity to find its way back to its place of origin. A good example of this is the Skylab space station which spent 6 years aloft, in orbit around the Earth. In July of 1979 it made a dramatic and much publicized reentry into the Earth's atmosphere. Pieces of Skylab were found scattered from Western Australia to the Indian Ocean.

More recently, in March 2007 a jet liner flying from Santiago, Chile to Auckland, New Zealand encountered a blazing fireball which appeared to break up into smaller fragments as the jet was approaching the east coast of New Zealand. The pilot estimated that the fireball was about 5 nautical miles from his aircraft. At first it was thought that the fireball might be the remnants of the Russian progress cargo ship that had been released by the International Space Station. The Russians, however, denied that this object was their ship which they said had reentered the Earth's atmosphere nearly 12 hours after the Chilean jet's encounter. Whatever the object was, it was later determined that only 20 seconds separated the jet from the fireball. In aviation terms this would be regarded as a close call. The crew of the jet stated that the fireball was so close that they could hear the roar of it. This incident along with Skylab and many other close calls over the years, including both space shuttle disasters have added a whole new list of suspects in the fireball phenomenon.

The 20th century has also brought us a phenomenon that at any earlier time in history would have been looked at with not only skepticism and disbelief but most likely heresy as well. It deals with the issue of Earth's uniqueness in the cosmos. Are we alone in the universe? This seems unlikely when one considers the billions of stars and galaxies that we know exist. The chance of Earth being the only planet that harbors life seems remote and almost statistically impossible when one considers the numbers of stars that are out there. It wasn't until the 1990's that astronomers were able to locate planets orbiting other stars in our galaxy. They were able to accomplish this by detecting slight wobbles of the star caused by the gravitational pull of what can only be an orbiting body. Before this time our solar system was the only known star system known to have planets. In the first half of the 20th century, mankind took to the skies and as I have already mentioned, changed the way we view the heavens. It was about this time that people began to ponder the possibilities of extraterrestrial life. If we could take to the skies, why couldn't some alien race from another star system? This led to another question; what if their civilization was more advanced than ours? Was it possible to travel from one star to another?

The UFO phenomena probably had its roots during the early stages of manned flight. In the late 1890s there was a rash of sightings. People reported seeing mysterious airships in the skies across the United States and Canada. However, it was at the beginning of the Cold War, after the 2nd World War that UFOs suddenly became a staple of mainstream culture. People had now found another suspect to throw into the fireball cauldron. Let us look at one of these reports.

On December 9, 1965 at approximately 4:43 P.M. a bright fireball was spotted over a large area near Lake Erie. It was estimated that the fireball had an apparent magnitude of—15 and was visible according to most eyewitnesses for only 3 or 4 seconds. Loud reports were heard directly after the visual observations which could have been sonic booms as the mysterious object passed the sound barrier. Witnesses also reported seeing a smoke trail which some say was visible for at least 30 minutes after the fireball was first seen. There was nothing special or even spectacular about this fireball which seemed to be an ordinary meteor burning up in Earth's atmosphere. In fact, in any other century this probably would have been the case. However, this was the year 1965. The Cold War was in full stride, and the Kennedy assassination was only a little more than two years in the past. The mood in the United States at this time was one of suspicion and cynicism. A meteor was just not an acceptable explanation for this extremely bright fireball. It had to have been something else. Perhaps something insidious or treacherous. This view was especially true to the citizens of Kecksburg, Pennsylvania. To this day, a good many residents of this town believe that something crashed in the woods only a mile outside of the town limits. One witness described the object as being of a "a burnt orange color, about 10 feet long and shaped like an acorn." . . ."It was smoldering, with sparks coming off it and smelled something like sulfur." Another witness identified it as being "one big, huge piece of metal buried in the mud, of a goldish, copperish yellow with marks on it similar to Egyptian hieroglyphics.

Whatever the object was that was seen in the woods outside of Kecksburg, some eyewitnesses believe that it was commandeered by the military that same night, being loaded onto a flatbed truck under an armed guard and furtively carried away to parts unknown. Ufologists have their theories as to what the object might have been. One theory is that the military carried off the wreckage of an alien spacecraft. A competing theory is that the acorn shaped object that was quickly hurried out of town that night was a captured Soviet space capsule. Many questions remain to be answered. For instance, what were the strange hieroglyphic-like markings seen on the side of the object? Could they have really been Cyrillic letters? Was the fireball seen over Ontario and most of the states bordering the Great Lakes even responsible for the object that was seen in Kecksburg?

Astronomers Von del Chamberlain and David Krause of Michigan State University believe the fireball exploded somewhere south of Windsor, Ontario. If this is correct, than the object seen at Kecksburg could not have been the same fireball seen over the Great Lakes on the evening of December 9, 1965, since over 300 miles separate the 2 locations.

A number of years ago the television program *Unsolved Mysteries* hosted by the late Robert Stack highlighted the Kecksburg mystery and gave the incident a broader national audience. A replica of the acorn shaped object rests on a platform at the Kecksburg volunteer fire department. The town is now a popular tourist attraction for UFO hunters, and others seeking an answer to a good mystery.

As I have now shown, fireballs have evolved from being objects which were dreaded, feared and of unknown origin to objects which were finally figured out to be nothing but space rocks made of the same elements that we find here on Earth. The 20th century, for all of its technological progress has ironically brought the fireball phenomenon full circle. For sure, most fireballs seen in the sky are of meteoric origin, but over the last century our mindsets have changed. We have become a world of cynics. No longer can the simple question "Did you see that fireball last night?" be asked without a second question being put to the person who has answered in the affirmative to the first question. The question being "So . . . what do you think it was? . . . A meteor? . . . An alien spacecraft? . . . A secret government weapon? . . . A spy satellite? . . . Space junk? Thus, the fireball question deepens . . . and is almost as enigmatic today as it was during the so-called Dark Ages.

Chapter 7

THE TUNGUSKA FIREBALL

It was a morning like any other in this land of permafrost. The region was sparsely populated by humans due to the inhospitable climate, but the reindeer herds went about as if business was usual. They had no way of knowing that this would be their last morning. If reindeer had the ability to reason like a human, they would have concluded that it was one beautiful morning in the taiga. Indeed, it was a crystal clear morning over the Stony Tunguska river in Central Siberia.

However, soon all of this was about to change. This remote region which lies just north of the 60th parallel was about to experience something that to this day still lies shrouded in mystery. Sometime just after 7:00 A.M. a bluish-white light appeared in the sky over Mongolia. It was shortly after spotted by herdsmen over Lake Baykal in Southern Siberia. A few minutes later the fireball crashed through the lower atmosphere with a deafening roar and exploded about 5 to 10 miles above the Earth's surface. The fireball could not have picked a better place to end its existence. The place of its termination just happened to be in one of the loneliest and secluded parts of our planet.

The date was June 30, 1908. Within seconds of the explosion hundreds of square miles of the Russian taiga had been leveled. Millions of trees were destroyed, their trunks scorched by the fire. Most of them lay uprooted , or bent over as if they had paid reverence to their cosmic visitor. It is estimated that the blast was the equivalent of 1000 Hiroshima bombs. If it had occurred anywhere else, the consequences would have been devastating. I have mentioned that the fireball struck just north of 60 degrees latitude. If it had struck just a few hours later in the day it might have leveled the cities of St. Petersburg, Helsinki or Oslo, all of which happen to lie on or close to the 60th parallel. Fortunately, as it turned out, besides the destruction of the forest the only fatalities happened to be the wildlife in the area. Herds of deer were killed along with birds, foxes, sable and bears. There is not one record of a human fatality. However, if some solitary hunter or lone trapper happened to be in the general area near ground zero, they would have met a terrible end.

Due to the isolated region in which the blast occurred, news of the disaster spread slowly— so slowly in fact that it would be nearly 20 years until a scientific team went in to investigate. This is not to say that it was not noticed at all. There were plenty of witnesses of the Tunguska fireball which had been reported by the local press. In the town of Irkutsk a reporter wrote in the newspaper *Sibir* that people in the small hamlet of Nizhne-Karelinskoye saw "a bluish-white light in the form of a pipe, high in the sky and moving downward. It was so bright that the eye could not behold it. After crashing into the forest a massive cloud of black smoke could be seen, and a forked tongue of flame came through the smoke." The townspeople feared the world was coming to an end. Amazingly, the town of Nizhne-Karelinskoye lay 150 miles from the blast site.

The newspaper *Krasnoyatets* reported a few weeks after the blast that people in the small village of Kenzma (which is about 130 miles from ground zero) heard a noise as if coming from a strong wind. This was followed by a loud crash and a subterranean shock which caused the buildings to tremble. The report went on to say that there were "50 or 60 loud bangs which sounded like artillery fire. Preceding the bangs a heavenly body of fiery appearance cut across the sky, and when the object hit the horizon a large flame shot up which cut the sky in two."

In mid-August the *Sibirshuya Zhizn* reported that an explosion on June 30 had frightening affects on some miners working 150 miles from ground zero. It says that "they all felt a trembling of the ground accompanied by a loud roar as though from thunder . . . the mine buildings creaked and groaned . . . people ran out in fear onto the street."

These were just a few of the reports that came out within weeks of the blast. However, these were all local reports. Outside of Siberia the Tunguska fireball was practically unknown. Although some talk of it reached Moscow and St. Petersburg the rest of the Western world would not hear of it for another 20 years. However, there were signs of the event that did not go unnoticed. Only a few days after the fireball struck the *London Times* wrote:

> The remarkable ruddy glows which have been seen on many nights lately have attracted much attention, and have been seen over an area extending as far as Berlin. One woman in England reported that after midnight on July 1 "the sky was so bright that it was possible to read large print indoors, and the hands of the clock in my room were quite distinct. An hour later, at about 1:30 A.M., the room was quite light as if it had been day; the light in the sky was then more dispersed and was a fainter yellow . . . it would be interesting if anyone would explain the cause of so unusual a sight."

Across the Atlantic the *New York Times* reported on July 3 that "remarkable lights were observed in the northern heavens on Tuesday and Wednesday nights, the bright diffused white and yellow illumination continuing throughout the night until it disappeared at dawn,"

Were these reports of strange nocturnal glows somehow related to what happened at Tunguska? At the time, of course, nothing was known of the blast, but some astute observers were able to see that something must have happened to cause these weird nocturnal lights. Some people had recalled seeing strange nocturnal glows after the 1883 eruption of the volcano on Krakatoa. Krakatoa is a small island in the Sunda Strait which is located between Java and Sumatra. It is part of the Indonesian Archipelago which extends for over 3000 miles and consists of over

17,000 islands. On August 26, 1883 the island was the scene of one of the loudest explosions and most violent volcanic eruptions ever heard or seen in recorded history. It was heard as far away as Australia which lies nearly 2000 miles away. The eruption blew a cloud of ash and hot steam nearly 5 miles high. Thousands of people were killed, mostly on account of the tsunamis that were generated from it. Others were killed or injured by the pyroclastic flow which annihilated many towns and villages.

The Krakatoa eruption also had global implications. It is estimated that average temperatures across the globe dropped by 1.2 degrees celsius. The volcano spewed a large volume of ash and gases into the atmosphere, including sulfur dioxide. This compound was caught in the winds of the upper atmosphere and quickly spread across the globe, eventually falling back to the earth in the form of acid rain.

A great amount of interest from the scientific community was generated toward Krakatoa. For a few days after the eruption, the skies around the world were darker than usual because of the great amount of ash caught in the atmosphere. However, this changed after a few days. Suddenly, the skies turned crimson and other bright colors. Scientists correctly ascertained that this was caused by light reflecting off of dust particles in the atmosphere. So it can be seen that when the strange nocturnal glows were witnessed by many during the early part of July, 1908, some automatically recalled the similarities from the eruption of Krakatoa a quarter of a century earlier.

Something had caused the strange glows during the early summer of 1908. But what? No major volcanic eruptions occurred at this time from where a source could be pinpointed. It was a puzzle that was to remain unsolved for another 20 years. One must remember the time in which the fireball struck. It was 1908, and imperial Russia had only recently emerged from a humiliating war with Japan that had left the Czar's government nearly destitute. Czar Nicholas II still lived the life of luxury in the imperial palace in Moscow while most of his subjects found it hard to put bread on the table. Needless to say, he was not a popular ruler. The time was ripe for revolution, but before that could happen a world war intervened which left the country in an even worse state. This left the government vulnerable and the revolutionaries took advantage of it. Maxim Gorky wrote in the newspaper *Novaya Zhizn* in 1917 "whenever they talk about it, everybody agrees that the Russian state is splitting all along its seams and falling apart like an old barge in a flood." The Bolsheviks eventually toppled the monarchy and sent Nicholas and his family into imprisonment before executing them in a remote mountain residence in the Ural's in 1918. The Red Army eventually went on to defeat the remaining supporters of Nicholas and the Civil War ended. It was a slow process, but eventually the destitute Russian economy began to reemerge, and the government (although it was now Communist) began to regain some stability.

Russia was now the Union of Soviet Socialist Republic, and another chapter in its history was about to begin. During the revolution, the government under the dictatorship of its leader Vladamir Lenin instituted the policy of "Prodrazvyorstka" which was a form of food apportionment. The government was to control food supplies. It was the first step toward Karl Marx's idealistic and utopian rule of the Proletariat. All industry and agriculture was to be controlled by the people's commissariat. However, Lenin underestimated the effectiveness of this policy. In fact, it was a total disaster. A black market soon emerged. The Russian farmers did not have any incentive to increase production. If the State was going to confiscate a set percentage of a farmer's crop why would the worker labor longer hours and get nothing in return? It seems that the Russian workers were not as idealistic and loyal to the revolution as Lenin and his Bolshevik followers had thought. The death knell of "Prodrazvyorstka" took place in 1921 when a great famine hit the region. A crop failure along the Volga river sent the new Communist government reeling to its knees and almost to the brink of another revolution. Joseph Stalin's biographer Nikolaus Basseches wrote "no one who was ever in that famine area, no one who saw those starving and brutalized people, will ever forget the spectacle. Cannibalism was common. The despairing people crept about, emaciated, like brown mummies . . ." It is estimated that over 5 million Russians died from the famine. Lenin and his government were forced to concede.

In the spring of 1921 the government abolished "Prodrazvyorstka" and instituted it's "Ekonomicheskaya Politika" or "The New Economic Policy (NEP). The NEP restored private ownership to farms and some business, but the government still held the lands to be part of the collective state. Gradually the economy began to recover, and the Bolsheviks had averted a possible power struggle. With the famine over, the government could now concentrate on more constructive things. The U.S.S.R would from this time on take a more proactive approach in the way they handled business. If they were going to become a world power as Lenin hoped, they would first have to develop their technology which was sorely outdated as the recent war with Japan had shown. Gorky wrote in the *Novaya Zhizn:*

> The free association for the development and dissemination of the positive sciences has been organized in Petrograd. The association is made up of the most talented and distinguished representatives of Russian science. These honorable persons intend to establish in Russia the scientific institute in commemoration of the 27th of February in commemoration of the birthday of our political freedom. The aim of the institute is to broaden and deepen the work of scientists along all the lines of the interests of man, society, nation and humanity. The foremost of these interests is the struggle for life against those sources of disease which undermine our health.

> Biology studies the phenomenon of life; Bacteriology explores the sources of contagious diseases; medicine strives to exterminate them; hygiene studies and points out those conditions in which man's resistance to disease increases.

> The Biologist, the Physician, and the Hygienist must know chemistry and make use of physics to the same extent as the Botanist who studies the life of plants and the Agronomist who, relying on the work of the botanist and the Geologist-Soil Specialist, works to increase the fertility of the soil and to raise its productivity.

All the sciences are closely connected with each other, and they
all represent the striving of human intellect and will to conquer
the grief, unhappiness, and suffering of our life.

Gorky had high hopes for the future of Russian science. So too did the
new Russian Academy of Sciences which began organizing exploration
parties to support the various fields of science. This was something not
done on a great scale since the days of Vitas Bering's Arctic exploration
in the 18th century. In 1921 the Academy decided to finance an expedition
tasked with locating meteorites which had fallen throughout the country.
The choice to lead this expedition fell on a man named Leonid Alexeivich
Kulik. Kulik was an expert on meteorites. He had worked as a mineralogist
at the Mineralogical Museum in St. Petersburg. His mission was a
daunting one. He was to collect as much information as he could about
the place a meteorite had fallen, and if possible gather eyewitness reports
and material evidence which could be taken back to St. Petersburg and
studied. If anyone was up for the task, that man was Kulik. He possessed a
whirlwind of energy, and set about doing his work in veritable earnestness
and pleasure. Shortly before the expedition was to set off on this quest,
Kulik got wind of a meteorite that had fallen in Siberia some 13 years
earlier. He was given a note which read:

About 8 A.M. in the middle of June, 1908 a huge meteorite is said
to have fallen in Tomsk, several sagenes from the railway line
near Filimonovo Junction and less than 11 versts from Kansk. Its
fall was accompanied by a frightful roar and a deafening crash,
which was heard more than 40 versts away. The passengers of
a train approaching the junction at the time were struck by the
unusual noise. The driver stopped the train and the passengers
poured out to examine the fallen object, but they were unable to
study the meteorite closely because it was red-hot. Later, when
it had cooled, various men from the junction and engineers from
the railway examined it, and probably dug round it. According
to these people, the meteorite was almost entirely buried in the
ground, and only the top of it protruded. It was a stone block,
whitish in colour, and as much as 6 cubic sagenes in size.

Kulik was fascinated by the note and must have read it a hundred times. However, little did he know at the time that this dusty report which was now 13 years old would dominate his scientific study for the rest of his life. Ironically, nearly all of the content of this report would prove to be untrue. It was mere poetic license from a journalist of a Siberian newspaper back in 1908. Kulik, however, had no way of knowing this. It was a cold tip, and to the benefit of science he acted upon it. Kulik would later find out that the only truth to this report was the fact that a train had stopped after hearing a loud noise, but the meteorite that was supposedly examined by train engineers and other locals was pure fabrication. Filimonovo Junction was nearly 400 miles from the Tunguska blast site. This is proof enough that the content of the note was not genuine. Regardless, Kulik was soon on a train bound for Siberia to investigate. It was not long before he realized that the information in the note was false. He gathered a number of reports and tips and soon realized that he was potentially on to something big. Whatever it was that happened in the early summer of 1908 occurred somewhere much farther north than Filimonovo Junction in a region around the Stony Tunguska river. From the reports he was also able to ascertain the direction of the object as it traveled across the sky. It was first seen in the south, and had traveled in a northerly direction. He was anxious to view the scene of the fall, but by the time he had arrived it was late in the season, and travel to the remote region would be extremely difficult. He had come unprepared for a long sojourn into the taiga. He therefore spent the remainder of his time collecting more eyewitness accounts, and gathering as much information as he could.

Kulik returned to St. Petersburg and reported his findings at meetings of the Russian Society of Lovers of World Knowledge. It would be 6 years before he would get the chance to return to the region. However, this interim of time was not wasted. He spent more time collecting reports and filing them away. In early 1924 he received a report from a geologist who was working for a museum in Krasnoyarsk. The geologist reported that a native Tungus named Il'ya Potapovich had been an eyewitness of the event from a remote trading station. He had an interesting story to tell. This man's brother lived on the

Chambre River and had experienced firsthand the awesome power of the Tunguska fireball. The report from this geologist, whose name was Sobolev, was as follows:

> Fifteen years ago his brother, who was a Tungus and could speak little Russian, lived on the river Chambre. One day a terrible explosion occurred, the force of which was so great that the forest was flattened for many versts along both banks of the river Chambre. His brother's hut was flattened to the ground, its roof was carried away by the wind, and most of his reindeer fled in fright. The noise deafened his brother and the shock caused him to suffer a long illness. In the flattened forest at one spot a pit was formed from which a stream flowed into the river Chambre. The Tunguska road had previously crossed this place, but it was now abandoned, because it was blocked, impassable, and moreover the place aroused terror among the Tungusi people. From the Podkamennaya Tunguska river to this place and back was a three day journey by reindeer. As Il'ya Potapovich told this story, he kept turning to his brother, who had endured all this. His brother grew animated, related something energetically in Tungusk language to Kartashov, striking the poles of his tent and the roof, and gesticulating in an attempt to show how his tent had been carried away.

Kulik was fascinated by this report. Two years after he received this letter, I.M. Suslov a Russian ethnographer interviewed the wife of this eyewitness. Her name was Akulina, and her husband had since died. She told Suslov that at the time of the explosion there followed a fierce wind that threw the whole family into the air. She and her husband had briefly lost consciousness, and when they woke up she stated that the whole forest was on fire, and many trees had fallen. Suslov also interviewed an elderly Tungus named Vasiley Okhchen who had been living with Akulina and her husband the day that the fireball struck. He said "that at the moment that the tent was torn away he had awakened and was thrown aside by a powerful jolt. The ground shook and a loud roaring was heard." He went on to say that "everything around was shrouded in smoke and fog from the burning falling trees, and the reindeer herd had disappeared."

The reports kept coming in, and with every one Kulik gleaned a little more knowledge of what had transpired on that day nearly 2 decades earlier. Of particular interest to him were the reports that had come in from places nearest the explosion site. The Evenki Camp of Il'ya Potapovich's brother had been a mere 25 miles from ground zero. About 40 miles away was the trading station of Vanovara. A trader named S.B. Semenov was sitting on his porch when he felt the Earth shake. Kulik was able to interview this important witness during his first expedition to the site in 1927. He was also interviewed by E.L. Krinov who was a member of Kulik's expedition in 1930. Semenov's testimony to Krinov is as follows:

> I don't remember the year exactly, but more than twenty years ago when the fallow land was being ploughed up I was sitting in the porch of the house at the trading station of Vanovara at breakfast time and looking towards the north. I had just raised my axe to hoop a cask when suddenly in the north above Vasily Il'yich Onkoul's Tunguska road, the sky was split in two, and high above the forest the whole northern part of the sky appeared to be covered with fire. At that moment I felt great heat as if my shirt had caught fire; this heat came from the north side. I wanted to pull off my shirt and throw it away, but at that moment there was a bang in the sky, and a mighty crash was heard. I was thrown onto the ground about three sajenes away from the porch and for a moment I lost consciousness. My wife ran out and carried me into the hut. The crash was followed by noise like stones falling from the sky, or guns firing. The earth trembled, and when I lay on the ground I covered my head because I was afraid that stones might hit it. At the moment when the sky opened, a hot wind, as from a cannon, blew past the huts from the north. It left its mark on the ground in the form of little paths, and damaged onion plants. Later, it turned out that many panes in the windows had been blown out , and the iron hasp in the door of the barn had been broken. When the fire appeared, I

saw Kosolapov, who was working near the window of the house, sit down on the ground, seize his head with both hands, then run into the hut.

In the winter of the same year, the Tungus Ivan Il'yich , came to me, and said: "why don't you look for gold on Lakura?" The forest there had been torn up by a storm, and the earth has fallen away. There was thick forest there but I don't know where it has been taken. A channel has been dug out, and along the sides you can see all kinds of stones. It's dry in the channel, there's no water; the birds come and peck the stones. Our flour store on Lakura was burned.

Semenov was a key witness to the destruction of the Tunguska forest. Of particular interest was his mention of the "hot wind." This is eerily similar to eyewitness accounts of the atomic bombs dropped on Hiroshima and Nagasaki in 1945. Semenov's daughter, Kosolapova, was also a witness to the Tunguska event. She was 19 years old at the time and was with a girlfriend gathering water from a spring when "the sky opened up in the north" and "fire poured out." The two women immediately ran for cover. As they were running, they heard loud bangs like gunshots and covered their heads for fear that stones might fall on their heads. Kosolapova said that she found her father lying unconscious near the barn, and she and her friend helped to carry him into the hut. She also stated that "the fire burned brighter than the sun, and the earth came sprinkling down from the roofs."

Another witness from Vanovara was a man named Kosalapov. He was pulling a nail from a window frame using some pliers when he felt a fierce heat that scorched his ears. Initially he thought that the roof had caught fire. He looked up and saw Semenov sitting on the porch, and asked him if he had seen anything. Semenov answered in the affirmative. He then went into the house and was about to start working on the floor when he heard a loud crash, and debris came down from the ceiling. A pane of glass broke and the stove's door flew off its hinges and landed on a nearby bed. He then heard a sound which sounded like thunder that gradually lessened as time passed.

In the village of Kezhma lived a political dissident named Naumenko. He gave a vivid description of the fireball as it passed across the sky. He said that "it was irregularly-shaped brilliantly white somewhat elongated mass . . . it was in the form of a small ball of cloud with a diameter far greater than the moon's, and it had no regular outlines."

Near the village of Zaimshaya a man named Kokorin was sailing in the Angara River. He tied his boat to a place along the riverbank and was climbing the slope when he noticed a "fiery red flame flying obliquely towards the earth from the north." He further stated that the object was about 3 times as big as the sun, but not as bright. He watched the flame disappear behind the mountains to the northwest after which he heard loud bangs that sounded like "continuous gunfire." It is interesting to note the similarities in the way people described the sound of the object. It was variously described as sounding like "gunfire" ,"artillery", or a "cannonade." It was also said to have sounded like "thunder." These were the most common descriptions given by eyewitnesses. A man named Sarychev who was washing wool on the banks of the Kan River said that it sounded like "subterranean rumbling." another man working in the fields described it as "a subterranean roar."

Kulik relished these reports and waited for the opportunity to return to the area of the fall. He finally got his wish when the President of the Academy of Sciences authorized him to mount an expedition to the region to study the site. Kulik and his assistant, Gyulikh, left St. Petersburg (Leningrad) in early February 1927 and journeyed by rail to the Siberian town of Taishet where they traded for horses and provisions that they needed for their journey into the bush. In mid-March they arrived at Kezhma where they received numerous reports and legends about the uninhabited region to the north. Kulik must have been anxious to set forth, but he had to gather more information before he was to proceed. From the all of the reports that he had read, and witnesses he had talked to, he deduced that the fall area was somewhere to the north of Vanovara, the place where 19 years earlier Semenov had been hurled from his porch. Vanovara was the last outpost north on Kulik's journey into the Siberian taiga. Beyond this small, almost invisible trading post lay an inhospitable region that was rarely traversed by man. Hunters and trappers were the only people brave enough to venture into this area. These were hardy folk who were accustomed to living off the land for long periods of time. Most of them were born and raised in the taiga. Kulik was a scientist who had no experience venturing off into the bush. He would need a guide who knew the land. It was March, and snow still covered much of the ground. This would make travel difficult at best, at the worst impossible. But this was not all that Kulik had to worry about. The forest was covered with thick underbrush which would severely hamper their attempt to reach the fall area. Also, if they did not start out soon they would be plagued by the mosquitoes that infested the region. One would think that this far north, only a few hundred miles south of the Arctic Circle mosquitoes would be the last thing to worry about. This was not the case. Beneath the winter snow was a layer of peat hundreds of feet thick. Years and years of ancient forest waste had turned parts of the region into swamp land. Naturally, of course, where there is a swamp there will be mosquitoes. Kulik was aware of the difficulties that faced him but was determined to face them head on.

It wasn't long before Kulik was able to find the guide he was looking for. It was none other than Il'ya Potapovich who eagerly accepted the position. The day after arriving in Vanovara, Kulik, Gyulikh, and Potapovich packed a couple of horses and headed into the lonely wilderness. They did not get far. The horses, loaded down with supplies found it hard to walk on the freshly fallen snow. Discouraged, the party was forced to turn back to Vanovara. The persistent Kulik immediately set to work to find a new way of getting to the fall site. In this respect he hired a reindeer herder named Okhchen to help him. The new plan was to follow a reindeer track as far north as they could. When the track veered in a different direction, they would have to blaze a trail of their own. This was no small task for a party as small as theirs. Fortunately, the new plan worked. The party set out in full stride, and 5 days after leaving Vanovara, Kulik and his 5 companions which now included Okhchen and Okhchen's wife and brother reached the Makirta River. It was here that Kulik first noticed physical signs of the nearly 20 year old fireball. The trees on the hilltops had mostly been toppled with their crowns facing toward the south. It appeared to Kulik as if a great wind had taken them down. The trees in the lower part of the valley still stood upright.

The small party continued north for a few more days reaching Shakrama Mountain and then the Khladni Ridge. It was at this point in the journey that Kulik first realized the extensive damage caused by the fireball. He was mesmerized by what he saw. One can only imagine what he was thinking as he looked out over the surrounding countryside from Khladni Ridge. For as far as the eye could see the forest had been completely leveled. Massive Larch trees that had survived hundreds of Siberian winters had been thrown down like match sticks. Only in the deep valleys, between ridges and hilltops was anything left standing of the primeval forest. A juvenile forest had started to emerge, but the devastation of 19 years earlier was plainly evident. Kulik took notes and determined that the destruction of the forest was especially prominent to the east of Khladni Ridge. To the north and south he could see bare patches of snow on the hillsides where the forest had been completely obliterated.

Eager to continue the journey Kulik urged his companions onward but was rebuffed by the reindeer herder Okhchen who had seen enough. He refused to go any further. Much to Kulik's chagrin Potapovich fell in line with Okhchen. The reason for the two guides' reluctance to continue is unclear. However, a few days earlier at the Makirta River, Okhchen had to be persuaded by Kulik to go on because of the herder's superstitious fear of what might lie ahead. It should be noted that the Tungus people at this time still held the beliefs of their ancient ancestors. They were Polytheists, and one of their gods was named "Ogdy" who was believed to be the god of light or fire. It was generally believed among their people that Ogdy was responsible for wreaking the destruction of the Tunguska forest in 1908. Somehow the Tungus had displeased this god of fire, and he had cast a fiery flame from the heavens to let them know it.

Despite Kulik's entreaties, the two guides insisted on turning back. Discouraged, Kulik had no choice but to relent, and the party returned to Vanovara. Once there he wasted no time hiring some more guides and within a few days was back in the bush. This time, however, he traveled a good portion of the way using a raft on the rivers Chambre and Kushmo. It was during this 2ⁿᵈ trip into the Tunguska region that he first became aware of the radial pattern of the destruction and believed that the center of the blast site was at a place he called "the great cauldron." It was here that he was convinced a large meteorite lay embedded somewhere deep in the swamp. He described this area in vivid detail in his journal:

> The centre of the fall is an area a few kilometers across on the watershed between the basins of the River Chunya and the Podkamennaya-Tunguska. It looks like a vast cauldron surrounded by an amphitheatre of ridges and isolated summits. In the south, at a tangent to this circle of hills, the River Kushmo flows from west to east, a tributary of the River Chambre that flows from the right into the Podkamennaya-Tunguska. In the cauldron there are hills, ridges, isolated summits and flat tundra, marsh, lakes and streams. The taiga, both in the cauldron and outside it, has been practically destroyed by being completely flattened. It lies in roughly parallel rows of bare trunks without twigs and tops, their upper ends turned away from the centre of the fall. This peculiar "fan" of flattened forest is seen remarkably well from the summits of the ridges and high ground that form the peripheral ring of the cauldron. Here and there, however, trunks of the taiga forest have remained standing, usually without bark or branches. Similarly, in places, small strips and groves of green trees have been preserved. These exceptions are rare, and in each case easily explained. All the former vegetation of the cauldron, of the surrounding mountains, and of a zone of several kilometers around them, bears characteristic traces of uniformly continuous scorching unlike the traces of an ordinary conflagration. Vestiges of bushes and moss remain on both the flattened and standing forest, on the summits and slopes of the hills, and in the tundra and isolated islands of dry land among the water-covered marsh. The area with scorch marks is estimated to be tens of kilometers across. Part of the central region tundra, covered with bushes and forest several

kilometers in diameter, of this "burnt" area bears what looks like the marks of lateral pressure. This has gathered it into flat folds with depressions a few meters deep, elongated roughly perpendicularly to the northeast. In addition, it is strewn with dozens of peculiar flat "holes" varying from several meters to tens of meters in diameter, and several meters also in depth. The sides of these "holes" are usually steep, although flat sides are also encountered; their base is flat, mossy, marshy and with occasional traces of a raised area in the center. At the northeast end of one of the sections of tundra the mossy layer seems to have been pushed aside for several meters from the foot of the hills and replaced by marsh. On the other side, at the southwest corner of the cauldron, the marsh ends in a chaotic accumulation of a mossy layer.

It was after a careful examination of "The Great Cauldron" that Kulik became convinced that not only was there a large meteorite embedded in the ground, but that it had broken up in pieces before impact. He believed that the proof were the small depressions or holes that he encountered during his initial survey of the area. However, on this first expedition he was not equipped with the necessary tools to prove his theory. He would need excavating equipment which he did not have. There was only one thing to do and that was to return to civilization. He left the Tunguska region and was soon on a train for St. Petersburg where he gave a his initial report to the Academy of Sciences.

Kulik was a determined man and convinced the Academy that he was on to something big. They therefore provided him with the resources for a 2nd expedition which set out in the early spring of 1928. This time there was a photographer named Strukov that would fall in with the party and chronicle the whole expedition. Strukov took the first photographs of the flattened forest, and was eventually able to make a film of it. Famously, at one point in the journey, he was taking pictures when Kulik's boat overturned in the rapids. His leg got caught in a mooring rope which, lucky for him, probably saved his life.

This 2nd expedition was much better equipped than the first. In order to prolong their stay they built cabins to live in, sheds to store their food, and equipment to protect it from wolves and bears. They made an attempt to dig trenches in the peat bogs hoping to find fragments of the meteorite, but they were unsuccessful due to the amount of water that kept pouring into the trenches. Kulik was dismayed to find that he did not have pumps large enough to remove all of the water. This was a severe blow to the 2nd expedition and prevented Kulik and his team from coming up with anything concrete in which they could take back to St. Petersburg for analysis. They did collect soil samples, but were disappointed that they were not able to find any evidence of a meteor. The one thing Kulik did conclude from this 2nd expedition was that the "holes" that they had seen the year before were not caused by meteor fragments. It was ascertained that they were natural indentations due to the permafrost.

The following year a 3rd expedition was launched. Once again Kulik was at the helm, but this time he had a host of scientists from different fields to assist him with his efforts. He was also better equipped with boring equipment, and pumps that enabled the team to remove the water from the trenches. They were also adequately supplied with provisions and material to survive the cold Siberian winter.

Kulik's team took more soil samples, and once again set to work in the trenches in an attempt to find evidence of the meteorite that Kulik still firmly believed lay immersed in the swamp somewhere. He was particularly fascinated with one gigantic impression in the swamp which he coined "The Suslov Hole." He focused on this spot which lay inside the "The Great Cauldron." The hole was more than 100 feet in diameter and seemed like the best bet where he could uncover some of the evidence that he was desperately looking for. After setting up their equipment, the team set to work excavating the hole. However, after weeks of hard labor, they were discouraged to find a tree stump near the center of it. This tree stump was the death knell for Kulik's theory of a large meteorite having been the cause of "The Suslov Hole." If a meteorite had been responsible for creating the hole, then surely the stump would have been blown to bits. Discouraged, but not defeated, Kulik was still determined to continue the search for meteoric fragments which he was positive were still somewhere in the area. Instead of giving up and returning to St. Petersburg in defeat he decided to concentrate his efforts on an area known as "The South Swamp." Kulik reasoned that if the meteorite did not land in "The Suslov Hole" than surely it must have landed here. Some of the locals had told Kulik that this area only became a swamp after the great fireball of 1908. This would have been a massive undertaking, and Kulik was running out of funds and time. After a cursory effort he was forced to concede, and left the Tunguska region with the hope that he would return the following season to investigate "The South Swamp" in greater detail. Unfortunately for him, a 4th journey to Tunguska would have to wait.

It would be nearly 10 years before Kulik had the opportunity to visit Tunguska again. He returned to "The South Swamp" buoyed by some aerial photography taken of the site the previous year. It was the last time that he would ever get the chance to find the cause of the Tunguska Event, the work that had consumed the better part of 2 decades of his life. Once again the expedition ended in disappointment for Kulik. When he left Tunguska for the last time, he was still convinced that it was a meteorite that had caused the devastation of the forest. On April 14, 1942 Leonid Alexeivich Kulik died of Typhus in a German prisoner of war camp. He was 58 years old. He had been serving selflessly as a nurse trying to save the lives of his incarcerated countrymen when he contacted the disease and succumbed. To this day no evidence of a meteorite has ever been found at the Tunguska site. In 1958 the Soviet Union issued a commemorative postage stamp in his honor. The stamp shows a bearded Kulik on one half, and the fireball that he will be forever linked with on the other.

Illustration 9

Since the days of Kulik there have been many other theories espoused as to the nature of the Tunguska fireball. Probably the most popular theory to arise was that the fireball was a comet. This theory was first proposed in 1930 by the British astronomer Francis Whipple who worked out of Kew Observatory in London. The internal composition of a comet was not known until just recently during the last passing of Halley's comet in 1986. A comet is made up of 3 distinct parts; a nucleus, coma, and tail. The nucleus is an amalgamation of ice, rock, gas and dust. The coma is the shroud of hot gas that surrounds the nucleus. The tail of a comet is made up of gas and dust and depending on the size of the nucleus can extend a million or so miles through space.

Whipple, and others called attention to the strange night glows seen across western Europe during the early summer of 1908. He believed it was possible that these glows could have been caused by the Earth moving through the doomed comet's tail. This is an intriguing speculation, but it is not the only possible evidence that may link Tunguska to a comet. Most of this evidence deals with the properties that are generally associated with a comet. If the comet was made up primarily of ice, most of it would have evaporated as the fireball plunged through the Earth's atmosphere. Keeping this in mind, it might explain why no impact crater has ever been found. However, some people have argued against the comet theory primarily on the lack of visual sightings. The Tunguska fireball was not seen until its approach over Mongolia a few minutes before exploding. Why wasn't it seen in the days and weeks leading up to June 30, 1908? Comets, when they appear, are usually seen for days or even months in the sky before they disappear from view. So if a comet was the culprit, why did it go unnoticed? Some scientists point to the angle of entry, or to the relatively small size of the comet. We can be sure of one thing. If it was a comet it must have been a small one. If Halley's Comet were to strike the Earth, it would produce devastating results around the globe. Life as we now know it would cease to exist, and the Earth would be cast into a period of darkness that would break the tenuous food chain.

The comet theory holds up well and is a likely candidate, but there are other theories that have been proposed in recent years that simply cannot be ignored. In the early 1970's two American scientists A.A. Jackson and Michael P. Ryan suggested that the Tunguska Event could have been caused by a mini black hole. They realized that no impact crater had ever been discovered which would have been a dead ringer for the meteorite theory. Also, because there has never been any meteoritic material found, they reasoned a meteorite could not have caused the destruction.

Black holes have only recently come to the attention of our knowledge of the universe. In fact, only in theory do they exist. No one has ever seen one, or for that matter detected one for certain, but it is generally believed that the universe abounds with them. In lay terms, one of the ways that a star ends its life cycle is by collapsing in on itself. The star material becomes so dense that nothing can escape from it, even light. Eventually this dense mass becomes so great that it begins sucking in all of the matter around it, including nearby stars, and some astronomers even speculate whole galaxies. Indeed, Some astronomers are certain that there is one of these ravenous star suckers in the center of our own galaxy. To give an example of how strong a black hole is, a black hole with the mass of a small planet like Mercury would have a diameter probably no larger than the eraser on the end of a pencil! However, if this minute speck of cosmic dust hit the Earth it could conceivably cause the damage seen at Tunguska.

Stephen Hawking, the British physicist famous for his book *A Brief History of Time* believed that it is possible that there are numerous mini black holes traveling through space. These black holes could have been created near the beginning of time during the incipient stages of our universe. However, whether or not they exist is still purely conjectural, but it is an enticing theory. One of these black holes could very well have been the invader of the Tunguska forest.

Supporters of the black hole theory point as evidence to the fact that Tunguska had no impact crater. A black hole would not leave one. However, detractors point out that if a black hole did strike the earth, it would have passed through its layers of crust, mantle and molten core and would have easily passed through the other side. Eyewitness reports and evidence at the Tunguska fall site have made it possible to roughly calculate the fireball's angle of entry. According to these calculations the black hole would have exited the earth somewhere in the north Atlantic Ocean. Obviously, if this had been the case, there would have been devastating evidence to confirm it. There would have been Tsunamis which would have caused much destruction to coastal cities along the Eastern North American seaboard and western Europe. There is no record or evidence of anything of the kind. In fact, microbaragraph records show normal levels during the time period in question. Since black holes are still not that well understood, we cannot discount this as being a possible reason for the Tunguska Event of 1908. That said, the way it appears at the present time, evidence does not seem to support it.

Probably the most controversial theory to arise in regard to the Tunguska Event is that it was caused by some sort of extraterrestrial vehicle. Yes, that's right . . . it has been proposed that a UFO crashed in the Siberian taiga on that early June morning back in 1908. This is a very interesting theory, but for obvious reasons holds little weight in the scientific community. However, it does have its supporters even if most of them hover around the fringe elements of science.

The first person to suggest this theory that an extraterrestrial space craft may have exploded over the Tunguska was a Russian writer named Alexander Kasantev. Kasantev had been an eyewitness to the aftermath of the carnage wrought at Hiroshima after the atomic bomb exploded over that city in 1945. He noticed the uncanny similarities between Tunguska and Hiroshima. Contrary to popular belief, the bomb exploded in the air at Hiroshima, not on the ground as most people would think. At ground zero, directly underneath the explosion was a pocket of space that appear to have suffered only minimal damage when compared to the total destruction of the rest of the city. Buildings stood upright, and some trees remained standing, their branches and crowns burned away from the intense heat, but somehow they had survived the blast. At Tunguska, minus the buildings of course, ground zero looked very much the same. There was an area estimated to have been directly under the blast where the trees stood erect. Like Hiroshima, their branches and crowns had been burned off. Kulik noted that they looked like telegraph poles, and this area came to be known as the telegraph pole forest.

To Kasantev, there was no question that the blast at Tunguska was nuclear in nature. The shock waves which were experienced by witnesses were similar to those experienced at Hiroshima and Nagasaki. A Russian aircraft designer named A.Y. Monotskov agreed with Kasantev. He used a computer model to show that the Tunguska object had slowed down considerably before exploding, something a comet or meteor could not possibly do. Supporters of Kasantev have also used eyewitness reports to show that the object changed course in its trajectory which is also something that a comet or meteor could not do. This is proof, they say, that the Tunguska fireball was guided by some kind of intelligence.

Nuclear physics was in its infancy in 1908. Einstein had only recently finished his General and Special Theories of Relativity which along with Quantum Mechanics would revolutionize the scientific world in the 20th century. It would still be many years after Tunguska when man would finally be able to harness the power of the atom. In 1908 it was still a generation away. Keeping this in mind, if the Tunguska fireball was nuclear in nature, then it would only stand to reason that it must have come form another world.

In recent years there have been other expeditions that have visited Tunguska. In October 2001 the BBC published a story entitled *Mystery Space Blast Solved*. It was written by BBC News online science editor Dr. David Whitehouse. It related that a team of Italian researchers had combined eyewitness accounts, seismic data, and a survey of the impact zone. Their conclusions strongly suggested that the object was a low density asteroid.

The most recent study of the Tunguska Event was undertaken by another Italian team and some of its conclusions were published in the on-line journal *Terra Nova* in April of 2007. The initial purpose of this expedition was to examine an area known as Lake Cheko which is located near the center of the blast area. The team was looking to analyze the sediment around the lake for possible clues which might shed some light on the origin of the now century old explosion. They ended up examining the lake itself, and decided that it is possible that the lake was actually formed by the impact of some cosmic body. According to this report, there is no hard evidence that the lake even existed before 1908. If this is the case, then it is possible that there is a large meteoric fragment buried somewhere beneath the lake. If this evidence is found then the century old mystery will be solved. Until then, the culprit remains an enigma.

As each year passes, vital clues disappear which may help to solve the mystery. A century after the Tunguska event the shroud of mystery is still firmly wrapped around this area of savage wilderness. The Tungus herders and trappers still work the rivers that their grandparents and great-grandparents did when the fireball struck, but the times have changed. Modern technology is slowly seeping into the region. The once small trading post of Vanovara is now a bustling little town with all of the modern amenities. Perhaps we will never know the true nature of the Tunguska fireball. We can only hope that one day the restless spirit of Leonid Kulik will emerge from the ancient permafrost and relate its secret to the mind of some inquisitive investigator.

Chapter 8

SIBERIA AGAIN . . .
THE SIKHOTE-ALIN FIREBALL

The Primorsky Krai is a sparsely populated mountainous region of eastern Russia that abounds with a variety of northern and southern animal species. A traveler traversing through its wilds is just as apt to run into a brown bear as he is a reindeer, or a tiger. The region is also rich in minerals. Coal and silver mines dot the landscape. Along with the fisheries and farming industry they give the Krai a fairly stable economy when compared to other parts of Russia. To the former Soviet Union the "Krai" as it was and is still sometimes known acted as an important strategic base for military operations. It shares its western border with China, and a small strip of its southern border with North Korea. The seaport of Vladivostok was at one time Russia's gateway to the Pacific. In a sense it still is but these days shares that honor with the port of Nakhodka which lies about 100 miles to the east. It is at Vladivostok that the Trans-Siberian Railway terminates at its most eastern point. From here it snakes northward through the Sikhote-Alin range before curving westward to its western terminus at Moscow some 6,000 miles away.

The Krai was the site of what many consider to be the most famous fireball of the 20th century. On the early morning of February 12, 1947 a very bright light appeared in the sky east of Lake Khanka which lies on the Chinese-Russian border. The light raced across the heavens from out of the northern sky. It was remarkably bright with a cigar-like shape, and left a trail of smoke and sparks in its wake. The fireball disappeared somewhere in the Sikhote-Alin range after some witnesses say it burst apart into many fragments. Shortly after this there was heard a loud noise which sounded to some like a heavy caliber artillery gun. This fireball is known to the world as the Sikhote-Alin meteorite, or the Sikhote-Alin meteor shower denoting the fact of its break up into thousands of pieces some 3 miles above the Earth. Unlike Tunguska, there is no question of what this fireball was. It was a relatively large meteor. Since that cold morning back in 1947, more than 9,000 iron meteorites have been found, which added up, weigh more than 30 tons. To this date it remains the largest recorded fall of a meteorite.

There are some questions which need to be answered in regard to this fireball. First, where did it come from? And second, is it likely to happen again? These are intriguing questions that need to be answered. As to the first question "where did it come from?" the obvious answer that any first grader could tell you is "outer space." But where in outer space? We do not know for sure, but it is highly probable that the Sikhote-Alin meteorite originated in the asteroid belt located between the orbits of Mars and Jupiter. There are millions of asteroids orbiting the sun in this region of space. The belt is estimated to be more than 150 million miles wide. At least 5,000 of these asteroids have a diameter of more than 9 miles. Some of them have even been mistaken for small planets. Indeed, the Sicilian monk-astronomer Giuseppe Piazzi thought that he had discovered a new planet when in December of 1800 he became the first to spot the asteroid Ceres which is the largest of these heavenly bodies. Ceres has a diameter of about 600 miles which is roughly half that of Pluto's. Ceres is large enough to have undergone the process of hydrostatic equilibrium, a process, when related to cosmic bodies, occurs when the body itself has sufficient gravity to overcome its rigidity. The end result is that the body takes on a spherical shape. This is one of the characteristics that distinguishes an interstellar object as being a planet, dwarf planet, or asteroid. Ceres has been classified as a dwarf planet because of its mass, but is still most commonly associated with the asteroid class because of its location. Another example of a dwarf planet would be Pluto which up until 2006 had been classified as the 9th planet in our solar system. Somehow astronomers demoted it because of its relatively small size and extremely eccentric orbit which sometimes takes it within the orbit of the gas giant Neptune.

We need not fear an Earth-Ceres collision. The orbit of Ceres has been known for years, and it remains in the vast region of space between Mars and Jupiter. However, much to our dismay we now know that there are asteroids that have orbits which bring them outside of the asteroid belt. In fact, there are some that have extremely long ellipses, and their orbits take them into the region of space in which the 4 inner planets orbit the sun. This brings us to the second question; "Is it likely to happen again?" In fact, most people would be astounded by the answer. It happens every day. That's right, every day thousands of rocks from space enter the Earth's atmosphere. However, most of them are no larger than a grain of sand and harmlessly burn up in the atmosphere. Some, however, are large enough to where they reach the Earth's surface as a meteorite. The Sikhote-Alin fireball was one of these. Needless to say, most meteorites are small and do not even leave an impact crater. A monster the size of Sikhote-Alin is a rare event. It is estimated that at least 70 tons of rock hit the Earth. This means that if 30 tons of it have been recovered, then there is at least 40 tons of this giant that has yet to be recovered.

The town of Iman which is now known as Dal'nerechensk lies about 40 miles from the Sikhote-Alin meteorite crater, or more properly put "craters" since the object broke up. At approximately 10:30 A.M. on this cold, icy morning an artist named P.I. Medvedev was at his home painting when by chance he just happened to gaze skyward and see the Sikhote-Alin fireball streaking across the sky at an odd angle, its multi-colored tail quivering behind it. He watched it until it disappeared near the horizon leaving him awestruck and with a subject for his next work. While the image of this extraordinary cosmic encounter was still fresh in his mind, he decided to put it to canvas. The result shows a colorful bolide streaking through the cloudy sky at about a 45 degree angle to the horizon leaving in its wake a trail of smoke. The bolide disappears behind some rolling hills in the background while some houses sit in peaceful repose in the foreground. The unseen inhabitants are oblivious to this intruder from the heavens.

Since no one was able to photograph the event, Medvedev's painting remains the only vivid glimpse of the Sikhote-Alin fireball. The original now resides in the Mineralogical Museum in Moscow, part of a permanent exhibition on meteors. In 1957, ten years after the fall, the Soviet government issued a postage stamp depicting Medvedev's painting to commemorate the event.

Medvedev may have been the only one to paint a visual for the outside world, but he was by no means the only witness of the fireball. At the small town of Paseka only a few miles from the area of the fall, a man named Kushnarev stated that he saw "a strong glare and a bright light which was brighter than the sun." He said that the fireball was high in the sky, and as it fell a thick black band of smoke descended steeply toward the ground. He then stated that "a cloud of black smoke seemed to rise from the ground," after which he heard an explosion that caused an earth tremor and shook the houses causing window panes to crack and break. Kushnarev said that he heard ten to twelve different explosions, and a noise that sounded like machine gun fire, which lasted for a good ten to fifteen minutes.

Illustration 10

Also at Paseka was a man named Taymanov who actually witnessed the fireball break apart. He described it as "white-hot fragments which flew steeply downwards in a dense group, fanning out as they fell . . . the larger fragments traveled in front of the others." Taymanov, along with Kushnarev and another man named Zaglyada, were three of the closest witnesses to the Sikhote-Alin fireball. Zaglyada who was also at Paseka and like Kushnarev, said that the fireball was as bright as the sun and high in the sky when he first saw it. He then said that the fireball broke apart leaving behind it a dark smoke trail with a reddish-pink tinge. He also stated that he heard loud explosions followed by earth tremors which caused the house to vibrate and window panes to crack.

Another witness to the Sikhote-Alin fireball was a man named Rodzyevsky who was farther away than the previous three witnesses, but still was able to give a vivid description. He first noticed a flash like lightning followed by the appearance of a fiery ball as bright as an electric arc. He stated that about 2 minutes after the fall he heard a loud explosion which shattered window panes. A few minutes later he heard smaller explosions which sounded very much like machine gun fire. Along with these small explosions he could hear now and then a loud explosion which finally tapered off into a roar like artillery fire. A column of grayish-white smoke could be seen above the explosion site and lingered in the sky for at least an hour afterwards.

There were other eyewitness accounts which occurred very near to the place of fall. A lumberjack named Ashlaban was stacking logs when he looked up into the sky and saw a fireball brighter than the sun streaking across the sky. This was the most common description of the Sikhote-Alin fireball.

Roy Gallant, a scientist and author of many books relating to space, was one of the first foreigners to visit the site of the Sikhote-Alin fall. Gallant got a personal tour of the site by the renowned Russian astronomer Valentin Tsvetkov during a visit to Siberia in 1995. He was also able to interview some eyewitnesses to the fireball. In the village of Meteority he interviewed Korney Shvets who happened to be working in a bakery at the time. Shvets saw blue flames sparkling in the sky with little fires trailing behind the main body. The windows in the bakery trembled, and a metal door of the oven flew open causing several hot charcoals to fall out on the floor.

An earlier researcher was E.L. Krinov who as I have previously mentioned was a member of Leonid Kuliks's expeditions to Tunguska in the late 1920's. Krinov, who died in 1984, was probably the world's leading authority on meteorites. He first visited the site of the Sikhote-Alin fall only a couple of months after it happened. He interviewed an eyewitness named Firtsikov who was a pilot, and standing by his aircraft in the village of Ulange which lies about 125 miles north-east of the fall site. He looked into the sky and saw a "fiery body in the northwest." The fireball was as big as the moon and was flying at an angle of about 60-70 degrees to the horizon. It disappeared behind the hills leaving a trail of smoke behind it. He then heard what sounded like loud bangs and a crash. Firtsikov and another pilot named Ageev flew to the town of Samarga, and found that the residents there had also seen the bolide. After spending the night in Samarga the two men arose the next morning and flew to Ulange, and on their way back happened to pass over the fall area. The ground was covered with snow, but the two pilots could plainly see impact craters which stood out because of the bright background of the snow. They reported their findings to the Geological Administration in Khabarovst on their arrival there a few days later.

The Geological Adminstration wasted no time in putting together an expedition. Firtsikov and Ageev agreed to carry a team of geologists to the fall site. The two planes flew over the area in order to allow team members a chance to observe the craters and their general layout from the air. However, because of its remote location, they were only able to land in Paseka about 7 miles away. The team consisted of three geologists: Yarmolyuk, Tatarinov and Onikhimovskiy. They set out on foot with supplies, but even though it was only a 7 mile hike, they made very little progress due to the deep snow. In fact, it took the men 3 days to reach the first of the craters. Yarmolyuk gave the following account of his initial impression of what he observed:

As we advanced, the proximity of the place of the meteorite fall became more and more obvious, for fragments of bed-rock hurled out by the meteorite, and snapped branches of trees, began to appear in the snow. Farther on fragments of rock weighing several tons of kilograms were found, and the snow which had been soft underfoot became dense, with a hard crust that supported the weight of a man. It was mixed with sand and clay, with large and small pieces of stone and forest debris. At last, in front of us appeared a huge crater . . .

Yarmolyuk and his two companions were the first people to explore the fall site. In the 60 years that has elapsed since then, thousands have visited this site. The type of people who make the trek into the blast site include scientists, treasure hunters and curious travelers whose only reason to visit might be to tell a good story to their grandchildren. There are even people who have a spiritual interest in the doomed cosmic visitor.

The same day that Yarmolyuk and his companions arrived at the site, a fourth geologist named Shipulin arrived from Vladivostock. The four men spent the day conducting a survey of the area. They located about 30 craters, and found some fragments which they quickly identified as being iron meteorites. The four geologists soon realized that they had found something unique. After a few more days surveying the area the expedition returned to civilization and reported what they had found to the Academy of Sciences. Another expedition was immediately put together. This one was jointly led by the chairman of the Meteorite Committee, V.G. Fesenkov, and Krinov whose experience in the field was considered indispensable. Shipulin showed the collected meteoritic fragments to Krinov who later described them as "having a very unusual appearance, and were more like shell or bomb fragments than meteorites." Krinov, however, soon dispelled any rumours that they might be the remnants of some American bomb gone astray. He analyzed the chemical and physical characteristics of the stones which were found to possess an amount of nickel content that was typically found in iron meteorites.

The expedition arrived on the site on April 27, 1947 which was a little more than 2 months after the fall. This was a big difference when compared to the two decades it took for scientists to reach the Tunguska site 20 years earlier. Krinov and his colleagues set to work mapping and surveying the area, and it wasn't long before they had found over 100 craters of varying size. The largest of these was about 80 feet across and 19 feet deep. They also found numerous fragments of the meteorite which ranged in size from meteoric dust flakes to rocks that weighed hundreds of pounds. The scientists also noted the affect that the fireball had on the surrounding forest. Hundreds of trees had been damaged or destroyed. Some of them had been completely knocked over, while some only had their branches torn off. In some respects there was an eerie resemblance of Tunguska. Near some of the larger craters the crowns of some of the trees had been lopped off, and they stood erect, branchless, like telegraph poles. One of the most fascinating finds came when Krinov found evidence of meteoritic fragments having passed straight through tree trunks. This was a stark reminder of the great speed and power that the fireball had possessed.

The expedition spent a month in the forest before returning to Moscow in June. Over the next few years the Academy of Science sent out more expeditions. During the 4th expedition in 1950 a meteorite was found which weighed a whopping 3800 pounds. This is the largest fragment of the Sikhote-Alin meteorite found to date.

It has now been over 60 years since the Sikhote-Alin fireball blazed across the eastern Siberian sky. Today, the region is still as remote as it was then. With each passing year the evidence of a cosmic impact diminishes. Craters can still be seen, but the trees that had been victimized by the blast have long ago rotted away. Perhaps a trunk or two remains in which a meteoritic fragment lay imbedded within its rings. There is an international market for meteorites which has grown up in recent years. The region has now become a hotbed for anyone willing to make the hike with a pick and shovel in an effort to possibly make a quick buck. The author of this book had only to make a quick search on E-Bay to find dozens of Sikhote-Alin meteorite fragments listed at auction. Are the listed items legitimate? . . . Who knows? Probably some of them.

Chapter 9

GREEN FIREBALLS OVER NEW MEXICO

During the late 1940s and early 1950s the State of New Mexico became a hotbed for fireball sightings. What was so peculiar about this episode in fireball history was not only the amount of sightings that were reported, but the bright green color that was associated with them. The fireballs attracted the attention of the U.S. government, and with good reason. Most of the sightings just so happened to take place around military installations or highly sensitive scientific training areas. New Mexico was and still is a haven for these places due to its desert and mountainous environment. In fact, it was at Alamogordo, New Mexico in 1945 that the first atomic bomb was tested.

Green fireballs are nothing new. They have been seen and reported by conscientious scribes for centuries. On August 13, 1887 J.N. Lockyer of the Royal Society of London "Observed in the northeast a magnificent fireball of emerald-green colour, as bright as Jupiter, with a very slow motion." In 1880 Henry Corder of the Royal Astronomical Society made some astute observations in an article he wrote for them. He stated "About 10 percent of all shooting stars show a distinct colour, the most usual being orange or red. The Taurids and other slow moving meteors seem rarely to get warmed above a red heat; the large ones, or those going a long way, often turn from orange to bluish-white like burning magnesium; sometimes the change is very sudden and startling. Green is a tolerably common colour, especially in slow moving fireballs about equal to Venus in lustre; they generally have a short train of red sparks." A green fireball was seen over New York State in 1828. It was described as being "Fine grass green with scintillations given off."

These three 19th century accounts show that the green fireball, though somewhat rare, is hardly unique. Two of these fireballs were also said to have been "slow moving" or "traveling with a very slow motion." This was a common theme shared with the green fireball frenzy that would hit the skies over New Mexico in the mid 20th century.

The New Mexico sightings happened to coincide with the UFO craze that hit the United States and other parts of the world in the late 1940s. It all started on June 24, 1947 when a pilot named Kenneth Arnold saw a group of flying disks near Mt. Rainier in the Cascade mountains of Washington State. Arnold described the disks as appearing like "flying saucers." The newspapers got a hold of his story and ran with it. In essence, what Arnold had started with this strange encounter was a whole new genre. That same summer of 1947 another celebrated case took place in Roswell, New Mexico. For those readers who are not familiar with it the "Roswell Incident" as it has become known to us is probably the most famous case of this new genre. Briefly summarized, something crashed into a rancher's field outside of Roswell in early July of 1947. An investigation conducted by local authorities along with the U.S. Air Force found that the debris in the field was nothing more than a downed weather balloon. This, of course, was the official version of what happened. However, over the years, and in light of new evidence, and previously unknown eyewitness accounts, a new explanation has emerged. It is now alleged by some that the Air Force covered up the incident. They believe that the debris strewn over the ranchers field was not a weather balloon at all but a flying saucer. To make the story even more interesting, witnesses have come forward and said that alien bodies were recovered from the wreckage. It is hard to say what the truth is. Numerous books and documentaries have come out which leaves this case as the most publicized incident in UFO history. The city of Roswell has become a haven for tourists, and has actually profited from this most celebrated of cases billing itself as the UFO capitol of the world. Due to the amount of content available on the internet and other media outlets I need not further expound on the "Roswell Incident" any more here.

The years immediately following the 2nd World War were years of suspicion and intense anxiety. It is easy to see how the disillusioned public thought that it was possible that our planet was the target of some hostile alien invasion. The world's two new superpowers; the United States and the Soviet Union, were engaging in an arms race. The threat of nuclear attack was not only real, but some thought imminent. It was during this time period that the green fireball drama began to unfold. The first sightings took place in November of 1948 when people reported seeing strange green lights in the sky around Albuquerque, New Mexico. These green lights were at first attributed as being nothing more than green flares. They were low on the horizon; furthermore, the descriptions given by eyewitnesses told officials at Kirtland AFB in Albuquerque that they were similar to the ones used by the military for training purposes. For a short while anyway, the fireballs were ignored by authorities. However, on December 5, 1948 an Air Force transport plane was flying close to Albuquerque at an altitude of about 18,000 feet when the crew noticed a bright green fireball in front of them. The crew reported that the fireball moved upward before leveling off. Meteors, though they do appear to sometimes shoot horizontally across the sky, usually descend at an angle close to 45 degrees to the horizon. To the crew of this airplane this did not seem to be a meteor. They reported the sighting to the tower at Kirtland.

The next report was even more dramatic and occurred on the same night that the transport plane had observed their fireball. A civilian plane flying close to Las Vegas, New Mexico reported that a fireball had almost collided with their plane. The captain of the plane stated that he at first thought the object was a shooting star. However, he soon realized that the flat trajectory of the fireball ruled this out. It first appeared as an orange-red color, but quickly turned to green as it approached their aircraft. The pilot was forced to take evasive action in order to avoid a collision. Due to the time and location of this report it is distinctly possible that the crew of this civilian plane saw the same fireball that the military transport plane had seen.

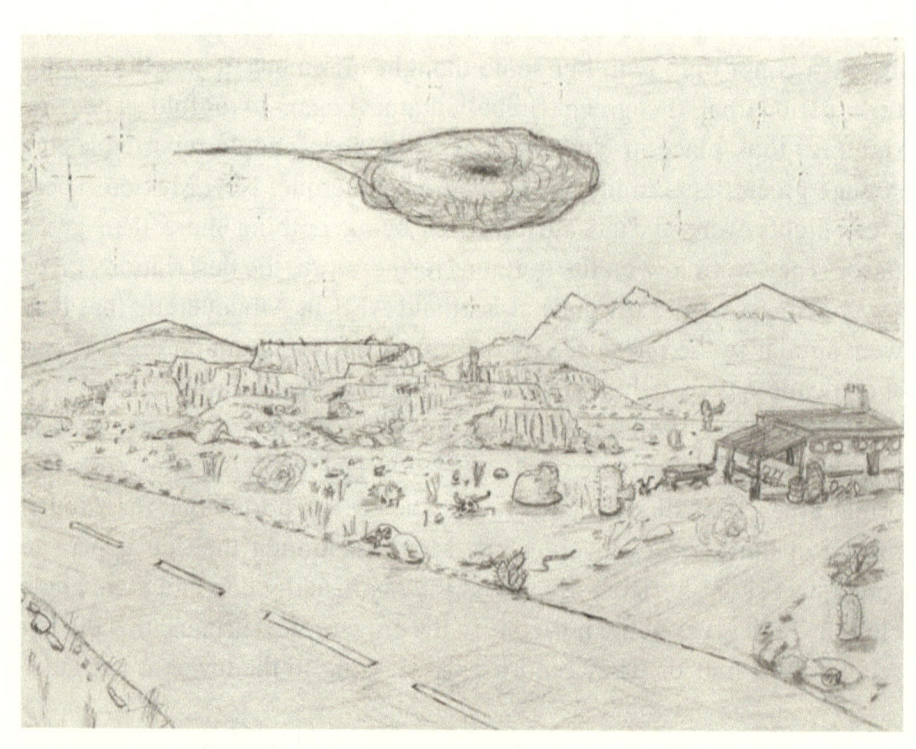

Illustration 11

After getting wind of these two reports the Air Force was finally convinced that something strange might be going on and that they should investigate the matter. Not only had one of these fireballs almost collided with a civilian aircraft, but they were also being spotted near Sandia Base and Los Alamos. Both of these places were nuclear research facilities. Undoubtedly this made Air Force officials very nervous. They called in Dr. Lincoln La Paz of the Institute of Meteorics at the University of New Mexico. La Paz was recognized as an expert on meteors, and had been working in the field for over 30 years. He had published numerous papers and treatises on the subject. If anyone could get to the bottom of this mystery, he was the man to do it.

La Paz read the reports and at first thought that they might be meteors. He had found meteorites before by using a unique method. He would gather eyewitness reports, and from them, get the general direction of the fireball. Using these reports he would plot a flight path which would lead him to the general location of the fall. He set out using this method in one case in an attempt to find what he was almost certain was a meteorite. After interviewing many people he soon realized that he was dealing with something on a grander scale. There were at least 8 different fireballs seen by people on the night of December 5 alone! After a cursory but diligent search he came up empty handed. No trace of a meteorite could be found at any of the possible fall sites.

The Air force was now truly concerned for the security of their military installations. The green fireballs had not abated in their invasion of the New Mexico sky. In fact the number of reports increased over the following weeks. They were spotted about every night, and not once was a meteorite ever found. Even Dr. La Paz was an eyewitness to one of these fireballs, and in his professional opinion it was not a meteor that he had seen. Throughout the months of December, 1948 and January, 1949 there was hardly a night that went by without a green fireball sighting. What bothered investigators the most was that this phenomena seemed to be unique to New Mexico.

There were plenty of theories that were floating around, but the one that bothered people the most was the theory that the green fireballs were some kind of new Soviet research instrument, perhaps even the prototype of some new missile system. Something had to be done. That much was certain, especially after Dr. La Paz suggested that the fireballs might be of artificial origin and not natural phenomena. On February 16, 1949 a conference was held at Los Alamos to discuss the green fireball phenomena. Attending this conference were a number of scientists including Dr. La Paz, a meteorologist Dr. Joseph Kaplan, and military intelligence officers. The purpose of this conference was to discuss the possible origin of the fireballs. The big question that everyone was dying to know was "What the hell were they?" After discussing the subject in great length it was decided unanimously that the green fireballs did indeed exist. There was no question about it. There were too many witnesses including people whose characters were beyond reproach. What the conference did not agree on was the origin of these fireballs. Were they artificial? or were they some kind of natural phenomena? Most of the attendees believed that they were some kind of natural phenomena like meteors. Lincoln La Paz and a few others disagreed with them. In an FBI office memorandum dated March 22, 1949 La Paz outlined his reasons for believing that the green fireballs were artificial. He cited one incident in which he had investigated. He called it "The Starvation Peak Incident." The reason that this fireball could not be classed as a meteorite fall was, La Paz believed, based on six things which he listed:

1. There was an initial bright light (no period of intensity increase) and constant intensity during the duration of the phenomenon.
2. Yellow green color about 5200 angstroms
3. Essentially horizontal path.
4. Trajectory traversed at constant angular velocity.
5. Duration about 2 seconds.
6. No accompanying noise.

However, despite his careful analysis and reasoning he could not sway the opinion of the rest of the conference attendees. They did, however, decide to try something. They would attempt to photograph the fireballs. From photographs they might be able to glean some more information. It was decided to employ the Air Forces Cambridge research laboratory to help solve this mystery.

"Project Twinkle" came into existence during the fall of 1949. The purpose of Twinkle was to photograph the fireballs by means of a theodolite, telescope, and camera. A theodolite is an instrument that can check the azimuth and angle of an object. If a fireball was photographed, it would be a very useful tool in determining the possible location of the fall.

While Project Twinkle was being organized and coordinated, there were also other means being employed to find a possible answer to the green fireball mystery. Dr. Kaplan, who had attended the Los Alamos conference, and whose advice was instrumental in the implementation of Project Twinkle, had interviewed witnesses to some of the fireballs. He showed each witness a color chart that delineated the wavelengths of the spectrum." He determined that each witness had chosen a wavelength in the yellow-green line of atomic oxygen. He was able to reproduce this radiation in his laboratory, and noted that it appeared as a bright emerald-color. He had achieved this color by way of a nitrogen-oxygen mixture. Kaplan sent these conclusions to General C.P. Cabell who was Director of Intelligence of the U.S. Air Force in Washington. He was convinced that the fireballs were some kind of new aurora that people were seeing, but that it was a natural phenomenon. Kaplan, however, could not explain the flat trajectories and low altitudes that were characteristic of the fireballs. He recommended that the Air, under the direction of Dr. La Paz, further investigate the phenomenon.

Dr. La Paz had other commitments, but Project Twinkle went into action in early 1950. After a little more than one year with no conclusive results it was decided that it was not worth pumping any new funds into the project. Nothing even resembling a green fireball had been photographed. Captain Edward J. Ruppelt, former director of Project Blue Book (The Air Force study on U.F.O's) said it best when he summed up the effectiveness, or rather the "lack of" of Project Twinkle. He said:

Of the three cameras that were planned for the project, only one was available. This one camera was continually being moved from place to place. If several reports came from a certain area, the camera crew would load up their equipment and move to that area, always arriving too late. Any duck hunter can tell you that this is the wrong tactic; if you want to shoot any ducks pick a good place and stay put, let the ducks come to you.

Project Twinkle was unceremoniously shut down. Any further sightings of green fireballs would be handled through the proper intelligence channels. The Government's reasoning had a lot to do with funds being allocated for the war in Korea which had dried up a lot of government spending on projects that were not deemed necessary for the country's defense. The final report of Project Twinkle listed a number of possible explanations for the green fireballs. As possibilities they listed balloons, airplanes, smoke rockets, and "small emissive clouds issuing from atomic installations." They also listed birds, planets and meteors in the list of possible suspects.

The report also stated that Dr. Fred Whipple, a well known astronomer from Harvard University, had studied the fireballs and surmised that small clouds when illuminated by the full moon could be the culprits. Whipple, along with Clyde Tombaugh (the discoverer of the recently demoted planet Pluto) had been working at the White Sands Proving Grounds and had spent countless nights looking up at the night sky, but had never seen any aerial objects that could not readily be explained. Perhaps Whipple had never seen anything strange in the night sky, but the report couldn't have been more wrong about Tombaugh. In 1948 Tombaugh was sitting in the backyard of his house in Las Cruces, New Mexico when he saw a "blue-green glow" fly overhead. He described it as being too fast for a plane and too slow for a meteor. The glow was oval shaped and appeared to be some kind of aircraft, for Tombaugh believed that he saw windows that gave off the same blue-green color only they were brighter than the rest of the lights on the aircraft. Whatever it was that Tombaugh saw he observed it just before the first green fireball reports started coming in.

After the demise of Project Twinkle the green fireballs seemed to vanish with it. Reports that had previously been coming in by droves now only trickled in a few at a time. In early 1952 the Air Force decided to step up their investigations into flying saucers and unexplained aerial phenomena. They created Project Blue Book headed by the rather reluctant Captain Ruppelt. Ruppelt was an aggressive investigator. Part of his job was to go through old case files and see if anything could be done with them. He began to ask questions, but was soon overwhelmed by the responses that he got. He talked with scientists and pilots at Los Alamos. Most of these people had at one time or another seen a green fireball but could not say for sure what they were. One pilot that Ruppelt interviewed gave a good analogy of a green fireball He said "take a softball and paint it with some kind of fluorescent paint that will glow a bright green in the dark, then have someone take the ball out about 100 feet in front of you and about 10 feet above you. Have him throw the ball right at your face, as hard as he can throw it. That's what a green fireball looks like."

The green fireballs gained a sort of notoriety when *Life Magazine* published an article in the April 7, 1952 issue which featured a young Marilyn Monroe on the cover. The headline reads, "There is a case for Interplanetary Saucers." The article itself details 10 eyewitness accounts that they label "incidents." The 10th one has to do with green fireballs and tells an astonishing tale that took place in Arizona on the night of November 2, 1951. A green ball of fire larger than the moon and many times brighter shot across the eastern sky in a straight line parallel with the ground. It finally exploded without making a sound. The article states that over 165 people witnessed this spectacle. A painting of a green fireball streaking through the sky among a background of dark foreboding mountains takes up much of the first two pages of the article. As one examines this painting one can feel the mystery that surrounds them. An enigmatic cloud of uncertainty haunts the viewer and is almost humbling. It asks you, "if the green fireballs are not natural, what are they?" Dr. La Paz did not have the answer. He had hoped that they would turn out to be some type of natural phenomenon, but his research had disagreed with this hope.

In 1948 the Office of Naval Research developed the Skyhook Balloon. This balloon was used primarily for research of the upper atmosphere and could reach an altitude of over 100,000 feet. The Skyhook is made of polyethylene, a light weight plastic material. These balloons at 100 feet, were larger than any that had previously been developed and deployed. They could also carry a payload of 300 pounds or better and stay aloft for 24 hours at a time. The first Skyhook launches took place at the University of Minnesota's airport, but it wasn't long before they were being launched from aircraft carriers and other places around the United States including the White Sands Proving Grounds in New Mexico. The green fireball phenomenon seemed to coincide with the launches of these balloons. Is it therefore possible that they were one and the same? This seems an unlikely scenario, but it cannot be ruled out. I have already stated that the Skyhook can reach an altitude of over 100,000 feet. Therefore, when it is dark at ground level, the sun's light is still reflected off of the balloon. Air currents at high altitudes can also explain the flat trajectories associated with the fireballs. The Skyhook Balloon has been known to travel at speeds exceeding 200 mph with these air currents. Keeping these things in mind, there is still one weak link in the Skyhook theory. This is the green color of the fireballs. There is nothing green about the Skyhook. However, taking this into consideration one must remember that it is possible that the green color is due to some optical illusion.

The Skyhook theory is only one of many that have surfaced over the more than half century since the green fireballs came into public view. They have also variously been attributed to test rockets, airplanes, and even helicopters just to name a few. Perhaps one of the strangest suspects that has been proposed was mentioned by British writer Desmond Leslie. In his classic collaboration with George Adamski entitled *The Flying Saucers Have Landed,* Leslie wrote of a Dr. Meade Layne who believed that the fireballs were "devices sent out to counteract radioactivity in the atmosphere, caused by the haphazard release of atomic energy." However, Leslie also wrote that Meade believed that these devices originated on the planet Venus! This book was published back in the early 1950s. This was years before satellite imagery and probes sent to Venus determined that the 2nd planet from the sun was a plane with 700 degree ground temperatures.

Instead of resembling some utopian world filled with glades and lush tropical forests as envisioned by early science fiction writers, it more resembled some renaissance image of hell.

The green fireballs eventually faded from public view, but not entirely. In the years since the New Mexico outbreak there have been scattered reports of green fireballs similar to the ones seen during the late 1940s and early 1950s. In fact, as I was writing the draft for this chapter, I was astonished to hear of a new report coming out of New Mexico. On September 13, 2007 at around 3:00 A.M. an emerald green fireball about four times the size of the full moon lit up the sky and was reported in places as far apart as Socorro and Santa Fe. It was captured on film by a camera at Sandia National Laboratory in Albuquerque. Dr. Lincoln La Paz had once commented to Captain Ruppelt that after the fireballs vanished from view no evidence or trace of them could ever be found. Over half a century later the mystery continues.

Chapter 10

SLOW MOVING FIREBALLS . . .
THE COMET

I can remember the night well. It was a crystal clear evening in a town appropriately called Crystal River on Florida's gulf coast. I was standing on a pier with my hands stuffed into the pockets of my denim jacket. It was a cool spring night in March. In fact, I remember the exact date, March 22, 1997. Comet Hale-Bopp was at the peak of its brilliance. I can remember thinking to myself "wow, this is nothing like Halley's Comet." Indeed, there was no comparison. I had observed the great comet named for its discoverer (of its known periodicity) eleven years earlier. I had been a senior in high school then and remember looking through the school's telescope at a fuzzy white patch and thinking to myself "what is all the hype about?" I was not impressed. The 1986 visit of Halley's Comet was a dud . . . a bitter disappointment for casual observers like myself. Comet Hale-Bopp was everything that Halley's Comet was not. It was bright . . . luminescent . . . . and downright spectacular. I observed this delightful visitor from the outer reaches of our solar system through my 4.5 inch Celestron Reflector. However, one did not need the aid of a telescope to take in the show. It was just as amazing without one.

Comet Hale-Bopp was first spotted nearly two years earlier by two amateur astronomers who found it independently, but at about the same time. The night that I stood on the pier at Crystal River happened to be the night that the comet reached its closest approach to Earth. The distance was only about 1.315 AU (Astronomical Units.) It would reach perihelion a little more than a week later and would be visible to the naked eye in some places up until the fall of 1997. Comet Hale-Bopp is now known as the "Great Comet of 1997." As of this writing, over ten years later, Comet Hale-Bopp is still being observed through deep-sky telescopes and has now traveled past the orbit of Uranus. The next time it reaches perihelion will be long after anyone reading this book has given up their mortal cares. Its anticipated arrival, in case anyone wants to mark it on their calendar, will be a little over 4,000 years from now.

Before proceeding any further in this chapter I would like to clarify its somewhat misleading title. There is absolutely nothing slow about a comet. They travel at great speeds through space, and I would be neglectful not to mention this to readers who do not possess a scientific background and are reading this book solely for its historical content. I call them "slow moving fireballs" only because that is our perception of them from our place here on Earth. As I mentioned in the last paragraph, Comet Hale-Bopp "was only about 1.315 AU" from the Earth on March 22, 1997. One astronomical unit equals the distance of the Earth from the Sun. This would be about 93 million miles. So adding the .315 to that amount gives you a goodly sum. The farther away an object is, the slower it is perceived. Although the stars in the heavens are moving through space at a very rapid rate they are so far away that it would take thousands of years to notice any change in the way that the constellations appear to us. So, with this possible confusion cleared up I would also like to mention one more thing before proceeding. I wanted to keep the subject of this book in as narrow a context as I could. Initially I hesitated before adding a chapter on comets, but I soon realized that I would be amiss if I did not mention them. After all, The Tunguska fireball could very well have been a comet. The same could be true for a number of other early fireball sightings that are either vague or misidentified because of the writers' ignorance on what he was recording. It was because of these assessments that I decided it would be appropriate to devote a chapter to these heavenly bodies.

Comets have been recorded since ancient times. In fact, the name "Comet" is a derivative of the Greek word "Kometes" which translated into English means "Hairy Star." In the ancient world there were many different theories that dealt with the nature of the comet. Aristotle believed that they were exhalations from the Earth that illuminated in the sky. His reason for believing this is not known, but he might have believed it had something to do with the make up of the universe. Aristotle believed that the universe was made up of five elements: fire, air, water, earth, and ether. Ether was supposedly the realm above the Earth where the heavenly bodies stayed fixed in their divine places. It is probable that he believed that since comets suddenly just appeared, and then suddenly vanished from the ether they could not possibly be part of the divine realm. Therefore, Aristotle probably reasoned that they originated on the mortal sphere.

Pliny the Elder, who wrote in the 1st century A.D., was one of the most prolific writers of the ancient world. He was a Roman, and like other writers of his time, wrote about things that he did not quite understand. He was aware of the works of the classical Greek scholars, and used their work as a base for his own philosophy. His *Natural History* is chock full of interesting, sometimes humorous and innocently amusing interpretations of the natural wonders of our planet. He wrote things as he saw them, but also recorded tales and anecdotes told to him by other people.

Pliny classified comets into groups based on their appearance, duration and movement. Meteors were also classed with comets, and Pliny only differentiated between the two by their physical characteristics. The first group Pliny called *Crinitae* which he described as being "shaggy with bloody locks and surrounded with bristles like hair." This horrible description reaffirms the ancient belief that comets were something to be feared and were a portend of something evil. The second group he called *Pogoniae* which appeared like "stars with a mane hanging down from their lower part like a long beard." He states that the third group *Acontiae*, "vibrate like a dart with a very quick motion." The fourth group *Xiphiae* are "short and pointed and pale, they shine like a sword and are without any rays." The fifth group *Discei* are "amber color in conformity with their name, and emit a few rays from their margin only." The sixth group is called *Phitheus* and are said to "exhibit the figure of a cask, appearing convex and emitting a smoky light." The seventh group *Cerastias* "has the appearance of a horn." The eighth group is called *Lampadias* and Pliny says that they are "like a burning torch." The ninth group *Hippias* is "like a horses mane; and possess a very rapid motion like a circle revolving on itself."

It is obvious by reading these descriptions that Pliny was not an eyewitness to most of them. He had gathered his information from other sources. A "Cerastias" he says was visible when the Greeks fought at Salamis. For those readers not familiar with this battle, the battle of Salamis was a naval engagement fought in the Mediterranean Sea between Persia and the Greek City States. It took place in 475 B.C. Since Pliny was writing in the late 1st century A.D., we can see that over 500 years had elapsed since this battle. It is unclear where Pliny derived this information. Herodotus does not mention a comet, and neither does Diodurus or Plutarch. He probably read it from a source that is now lost. It is possible that Pliny traveled to the city of Alexandria in Egypt where a great library is known to have existed. Indeed, it is known that Pliny had traveled to the African continent, but whether or not he found his way to Alexandria is not recorded.

Out of the nine types of comets that Pliny mentions by name, at least two of them are fireballs of the meteor kind. We know this because there is absolutely nothing "rapid" or "quick" about how we perceive a comet. The two in question would be the *Acontiae* and the *Hippias*. The *Acontiae* he said "vibrate like a dart with a very quick motion." This was almost certainly referring to a fireball of some sort in the fashion of a meteor. Pliny adds that the "Emperor Titus had described one in his very excellent poem as having seen in his fifth consulship; and that this was the last of these bodies which has been observed." Although this is not really important to our discussion on fireballs the Emperor that Pliny is calling Titus is not that Emperor who reigned from 79-81 A.D. and who usually goes by that name. He is referring to the Emperor Vespasian who was Titus' father. His fifth consulship would have been in the year 73 A.D. If this was the case, then Pliny would have been recording a very recent event. However, it is still unclear if, in this laconic passage, Vespasian had witnessed the fireball himself or if he had written the poem based on other eyewitness reports. The most intriguing part of this report is the last part of the sentence which says "that was the last of these bodies which has been observed." What does this mean besides the obvious? It seems to imply that this fireball was no ordinary shooting star, but rather some spectacular event. It must have been quite a sight, or left an indelible impression for Vespasian to have mentioned it in a poem.

The *Hippias* is described as "being like a horse's mane; a very rapid motion like a circle revolving on itself." This supposed Comet needs to be analyzed further. It is obviously not a comet. In the days of Pliny no comet could have been mistaken for possessing rapid motion. We then have to consider whether or not Pliny is describing a meteor. A meteor is definitely known for its rapid motion. However, does it look like a horse's mane? Perhaps a comet with its tail streaking behind it might be construed for a horse's mane, but we have already determined that Pliny could not be talking about a comet because of the rapid motion. So what then could a *Hippias* have been? The clue could be in the last part of the description which as you recall says "like a circle revolving on itself." Is it possible that Pliny was describing some kind of primitive aircraft? A balloon, I suppose, when looking at it from a distance with its teardrop shape could possibly be mistaken for appearing as a horse's mane. However, it would take a little bit of cajoling to get me to believe it. We must remember, however, that hot air balloons did not make their appearance until the French experiments in the late 18th century.

A Greek inventor named Hero of Alexandria who lived in the 1st century B.C. invented a device called the aeolipile. The aeolipile was a sort of cylindrical chamber which pivoted on a suspended shaft. Underneath the device was a fire which caused the chamber to spin and expel steam from two tubes protruding from it. This device is based on the principal of rocketry. No practical use for this device is known to have existed in Hero's time, but how do we know this for sure? Is it possible that some sort of steam rocket was invented 1000 years before the Chinese developed the first solid fuel rocket?

It is not known where Pliny gathered his description of the *Hippias* comet. Did he see one of these himself or did he get his information from a second hand source? We will probably never know for sure. Like the *Acontiae* comet, it is cloaked in mystery.

Pliny also mentions a couple of other types of comets that he fails to assign names. One of them he says is "a white comet with silver hair, so brilliant that it can scarcely be looked at, exhibiting, as it were, the aspect of the deity in human form." To a contemporary of Pliny this comet, or whatever it was, must have been an awesome if not terrifying sight to behold. Pliny has likened this stellar body with a human visage and given it almost sage-like qualities which cast it in a divine light. The other comet that Pliny describes is "shaggy, having the appearance of a fleece, surrounded by a kind of crown. There was one, where the appearance of a mane was changed into that of a spear." The element of war can plainly be read into this description. A "mane" changing into a "spear" and a "crown" are all objects that relate to war and empire. Indeed, throughout history war and empire are common motifs associated with comets. As I have already mentioned earlier, a comet was seen before the battle of Salamis in 475 B.C. In this case Pliny does not say if this comet was seen as an omen that portended some disaster, or a sign of a favorable event. To Xerxes and the Persians there was no question that the comet was looked on with fear when they realized that the battle was not going in their favor.

Probably the most famous example of a comet's appearance leading to war or battle occurred in the year 1066. In the spring of that year Halley's Comet appeared on its 76 year periodic orbit of the sun. 1066 was the year Duke William of Normandy crossed the English Channel in a fleet of ships and invaded the Anglo-Saxon kingdom of King Harold Godwinson. Only three weeks earlier the English army had fought a high pitched battle against a Scandinavian army at Stamford Bridge in the north of England. The English won that battle, but the fatigued army had no time to celebrate their victory. Harold gathered his men and headed south to meet the new threat lurking on the southern coast. By the time the English army arrived on the field at Hastings, the Normans had already gotten a foothold, and

William had time to make battle plans. The Normans soundly defeated the English, and William became King William I. He is known today by his better known alias "William the Conqueror."

The comet of 1066 is mentioned in the *Anglo-Saxon Chronicle*. It is also recorded on the famous Bayeux Tapestry. The Bayeux Tapestry is an amazing work of art 70 meters long depicting the Norman invasion of England. In one scene of the Tapestry it shows the newly crowned King Harold who has just been given the keys to the kingdom after the death of his predecessor Edward the Confessor. People are depicted in this scene gathering around their new king who is holding the orb and sceptre that symbolizes his stead. Harold seems to be firmly ensconced on his throne while the spectators crowd Westminster Abbey to pay him homage. Some are pointing to their new king while others are clapping. However, the mood of the tapestry changes in the next two scenes. We now see a group of people pointing up at the sky. One man looks back at his fellow stargazers holding a finger up at the heavens. He seems to be asking them a question or two. "What is this? . . . what can it mean?" At first glance the meaning is not clear and somewhat ambiguous. The viewer then turns to the object of interest. It is a fireball of some sort which is visible in the upper border of the tapestry. Upon closer examination it seems that the fireball is woven into the fabric in three distinct parts. A comet, of course, is made up of three distinct parts: the head, coma and tail. The man is obviously pointing to a comet. However, if there is still any doubt as to the nature of this heavenly body further evidence is woven into the tapestry directly above the stargazers and just to the left of the comet. The evidence comes in the form of words "ISTI MIRANT STELLA." This Latin translated into English means "These men wonder at the star." The star, of course, being none other than Halley's Comet. It would have been visible to the naked eye at around the time of Harold's coronation.

The next scene shows what appears to be a court page or some kind of messenger whispering something into Harold's ear. Harold seems to be listening closely and with a keen interest in whatever is being said. This time, however, there is no Latin inscription that tells us what it is. We can only make an educated guess. Two possibilities arise. The first is that he is receiving news of Duke William's displeasure at the recent turn of events. To William and his Norman kin, Harold was nothing more than an imposter. He was a usurper who had stolen the crown of England from its rightful heir, Duke William himself. William believed that the crown was his. An earlier scene of the tapestry supposedly shows Harold taking an oath in support of William's claim. In reality Harold had been shipwrecked off the coast of Ponthieu and taken prisoner by one of William's allies who turned him over to the Norman Duke. If some kind of oath was given to William by Harold, as the tapestry suggests, than it would have been most probably given under coerced circumstances.

The second possibility, and the one that makes the most sense, is that the messenger is giving Harold news of the comet. A ball of fire has been seen in the heavens, and for Harold this is not good news. God is obviously displeased with Harold and has sent this ominous sign as a warning. Harold appears disturbed by the news. In just three scenes of the tapestry, the mood has changed three times. It has gone from joy, which signifies a new beginning, to wonderment at a strange celestial body, to worry and possibly fear after the wonderment has been interpreted. There is no record of William's reaction to this comet. It is not delineated on the tapestry. One can only wonder and speculate. If he thought that it was a bad omen, it is unlikely that he would have attempted to cross the English Channel. As the old cliché goes, "actions speak louder than words," or for that matter, silence. In William's case either the comet meant absolutely nothing to him, or he felt that it was a sign from God directed at him to retrieve what he believed rightfully belonged to him. He had the Pope's blessing, and now, he might have figured he had God's as well! It is therefore possible that Harold was doomed by the appearance of a comet, and the whole course of British history was changed forever. This, of course, is only conjecture, but comets do have a way of affecting the way people behave.

As I have already mentioned, comets have been observed and recorded since ancient times. One of the earliest records that we have of a comet was recorded by Pliny. In 466 B.C. a comet appeared in the sky over Greece. At the same time at Egospotamos in Thrace, it is recorded that a number of rocks fell from the sky. Coincidence? Or was there something more? Did the Earth pass through the tail of a comet causing this massive discharge of rocks from the sky. The 19th century astronomer, Daniel Kirkwood, believed that there was an effort made to recover rocks from this ancient fall, but nothing conclusive came from the search. Of course, after 2500 years a lot of the evidence would have been buried or at least hard to discern due to chemical factors. From this early event we can see that comets and some fireballs are intertwined, and that the correlation between the two was probably known even during the days of Pliny. It is probably one of the reasons that comets were such dreaded sights to behold. It is now known that meteor showers occur when the Earth passes through the field of debris left by a passing comet. In those days, however, it appeared as if the gods were angry and decided to let their mortal subjects know it by hurling these fiery lights down upon them.

Probably one of the most famous comets in ancient history took place when one appeared over Rome in 44 B.C. The Roman Emperor Julius Caesar had just been assassinated and his successor, his stepson Octavius, had inherited the throne. Octavius took a new name, Caesar Augustus, and in honor of his stepfather decided that some games should be held in his memory. This was about the time that the comet appeared. According to the Roman historian, Suetonius, it appeared on the first day of the games at about an hour before sunset and was seen in the sky for a full seven days. Because of its timing, the comet was thought to possess the soul of the recently departed emperor, and was the vehicle that would carry it to heaven. Not being aware of the true nature of comets at this period in history, it is easy to imagine the public believing this. Julius Caesar was the most powerful figure of his day, and indeed, even today remains one of the most popular figures in world history.

Comets continued to awe and frighten the unsuspecting spectators well into the middle ages and beyond. The French surgeon and chronicler of prodigies, Ambrose Pare, writing in the 16th century, promulgated this fear of comets in his work *Des Monstres et Prodiges*. Pare writes "comets have never appeared without producing some bad effect, and leaving behind a sinister outcome." He mentions a "horrible and frightening" comet which appeared in the sky over Westrie in October 1528. This comet, he says, "was covered with blood, and was so terrifying that some people died looking at it." He adds that the comet lasted for an hour and fifteen minutes and appeared in the direction of the rising sun. This description would have shocked the readers in itself, but Pare was not through with his description. He wrote that "a curved arm can be seen in the comet brandishing a large sword which appears as if it is ready to strike. Near the point of the sword there are three stars, the one at the point lighter than the other two. The sides of the comet are covered with knives, swords, and other instruments of death along with hideous human faces, with their beards and hair bristling." Pare claims to have borrowed this vivid description from the writer Boistuau, who wrote about it in his *Histoires Prodigieuses*. Boistuau was supposed to have copied it from the work of Conrad Lycosthenes another 16th century chronicler of prodigies. The description of this renaissance era comet sums up best how our ancestors viewed these heavenly bodies. After the invention of the telescope, Copernicus' and Kepler's revolutionary discoveries brought about the period known as "The Age of Enlightenment." Our terror of comets diminished somewhat. However, it was the great British astronomer Sir Edmund Halley who finally put all the superstitious notions about comets to rest for good. Halley was the first to figure out that some comets orbited the sun at regular intervals. These comets are known as Periodic Comets. He was able to prove by mathematical calculations that the great comets of 1531, 1607 and 1682 were actually the same comet. He made a bold prediction that this comet would return again in 1757 or 1758. Unfortunately, for Halley, he did not live to see his prediction come true.

Today, we possess a considerable amount of information regarding the makeup and nature of a comet. This is due mostly to the new technology that we have developed in the last few decades. In March of 1986 Halley's Comet was at its peak magnitude to earthly viewers. The Giotto spacecraft had been sent up with a payload that contained a camera and was able to take stunning photographs of Halley's nucleus at a distance of only 600 Kilometers. The images that were received back on Earth surprisingly showed the nucleus to be very dark, and therefore much larger than previously thought.

We have now found hundreds of comets with a known periodicity. Comet Halley visits us regularly every 76 years, while Comet Hale-Bopp makes a grueling 4,000 year voyage around the sun. There are, however, comets out there in deep space that we do not even know about. In the early 1950s the Dutch astronomer Jan Oort theorized that a cloud of comets numbering in the billions existed within the gravitational attraction of our sun. This region of space, however, was well beyond the orbit of Pluto. In fact it extended over 100,000 astronomical units from the sun. This is an amazing distance when one considers that Pluto, as far away from the sun as it is, is only 40 astronomical units away. This definitely puts everything in perspective. If there are billions of comets lingering about in the vast reaches of space outside of Pluto's orbit, what do we know about them? The answer is, unfortunately . . . not much. Some astronomers theorize that some of these comets are occasionally knocked out of their orbits by collisions or the influence of nearby stars. One can only imagine some capricious wobble of a nearby star sending a flock of these icy bodies into the inner reaches of our solar system. It would thus appear that the fate of our planet just might one day suffer an ignominious pockmark by one of these truly ancient cosmic bodies. Let us hope that it is a long time from the present day, and that Ambrose Pare wasn't something of a prophet when he disseminated his tidings of fear nearly four centuries ago.

Chapter 11

THE LIGHTNING FIREBALL (BALL LIGHTNING)

In early September of 1998 my wife and I decided to take a road trip from my adopted State of North Carolina to Vermont. Along the way we stopped at many historical sites which dotted the landscape. One of these stops was at Saratoga Springs in upstate New York. We arrived late on the evening of September 6th and searched high and wide for a hotel. Everywhere we looked we were dismayed to find the ubiquitous "NO VACANCY" sign lit up in all its neon glory. We were about to despair and resign ourselves to having to spend the night in the car when we found a small motel with only one word lit, "VACANCY." I thought that perhaps the neon letters "NO" had flamed out years ago, especially after getting a second look at the place. In fact, I was almost convinced of it! It was a seedy looking dump with one of those signs in the window that said "COLOR TV." Are you serious? Color TV? That will get me to stop every time.

Needless to say, it was not this thirty year old antiquated sign that had me pulling into their parking lot. It was the lack of one of our primary needs, sleep, that had me turning the doorknob of the front office.

"You are lucky folks" said an old man sitting behind the front desk. "We have one room left."

I paid the old man cash since that was the only payment he would take, and he handed me a key to a room. "Have a good nights rest." he said in a way that reminded me of Norman Bates. "I'm sure I will" I remember thinking as a sickening vision of cockroaches and bedbugs passed through my head.

The next morning we were going to pay a visit to the Saratoga battlefield a few miles down the road. At that moment, however, after being on the road all day, we could think of nothing but sleep. It had been a nice drive through Pennsylvania and New York, and we had not run into any foul weather along the way. I did notice a few drops of rain on the windshield as I had pulled into the motel, but I did not think any more about it until after we had turned in and the sky decided to open up. Soon we could hear the rumbling of thunder in the distance, and see the flashes of lightning through the cheap curtain covering the window in our room. Slowly the thunder got louder and the flashes of lightning more intense, and we could almost believe we were on the western front during the 1st World War. It was one of the most violent storms that I had ever experienced with strong wind gusts added to the fury. After a while the storm abated somewhat, and the rumbling became more distant. Just as I thought that I might get a little bit of sleep, I was drawn to an illuminating glow from the window. Curious, I got up and parted the curtain slightly and found myself looking at something across the street that had me a bit perplexed. I can best describe it as a globe of light, almost too bright to look at directly. It appeared to be the size and shape of a basketball, but I must admit that I could have been deceived by the distance. The ball of light was pulsating as if it were attempting to increase in size. It was also moving in a small arc toward the sky. It was a good twenty feet or so off the ground. After watching it for ten or fifteen seconds I called my wife, who was just getting up, when suddenly the flaming ball exploded in a blinding flash which caused me to temporarily lose my eyesight. It was sort of like looking at a welders torch magnified to an even higher level. After slowly regaining my vision I began to ponder the cause of this strange light. I believed at first that it was a transformer which had been arcing and then exploded. However, even indoors with the door shut and the windows closed I still should have been able to hear the loud pop from one of these exploding. Working out on the railroad I have been witness to a number of transformer explosions over the years, and believe me, they have a distinct sound to them! I also searched the grounds the next morning to see if this could still be possible, but found no transformer in the area where I had viewed the fireball. So if it wasn't a transformer, what might it have been? I could think of no other explanation than that the mysterious globe of light that I had seen on the evening of September 6, 1998 was the rare and enigmatic phenomenon known as ball lightning.

What is ball lightning?

This is a good question, and I ask it just as much for myself as for the reader. At this point in time no one can say for certain what causes it. Indeed, it has only been in the last few years that science has even recognized this phenomenon. In the not too remote past reports of ball lightning were brushed off by most meteorologists as having no factual basis. It was generally believed that people were misidentifying some other phenomena. However, over the last few decades the realization that ball lightning actually exists can no longer be denied. There have been too many credible eyewitnesses to deny its existence. Over the past few years there have been attempts made to solve its mysterious nature.

A recent National Geographic article suggests a few possibilities. One of them is that plasma clouds are made up of charged particles that form new atoms. These charged plasma bodies emit a light that we see as ball lightning. Another competing theory suggests that traditional lightning may trigger the phenomena. This makes sense seeing that most cases of ball lightning seem to occur during or just after a thunderstorm. This theory holds that after a lightning strike, a vapor forms which then condenses into particles which amalgamate with oxygen. This oxygen rich environment then burns away. The result being a massive amount of electrical energy which has created a chemical reaction. This causes the glow that we see in ball lightning.

Eyewitness accounts of ball lightning vary when it comes to the size of the body and the duration of its existence. Generally, ball lightning is spherical with a small diameter, but descriptions are different. One thing is certain, however, and that is that ball lightning has been seen since ancient times. One of the oldest reports is recorded by the Greek historian Plutarch. It involves the Corinthian General, Timoleon, who is perhaps best remembered for defeating the Carthaginian army at Sicily on more than one occasion during the 4th century B.C. In 344 B.C. Timoleon and a small contingent of Greek soldiers and mercenaries set sail from Corinth and were on their way to Sicily when they encountered a strong wind. The whole fleet numbered only 10 ships, and was in deep water when, according to Plutarch, the sky opened up and a bright spreading flame issued forth from it. After a short time the flame was seen to form into a torch and began to hover over Timoleon's flagship where it stayed until it guided the fleet by its light to the safety of the Italian shore.

It is difficult to analyze a report like this. This is especially true of one that is 2,400 years old and has almost surely has been tainted by myth and legend. Timoleon and his crew saw something, but whether or not it actually guided their fleet to safety is debatable. We also must consider that Plutarch, who related the story, was writing about it after a lapse of 450 years. Taking all of this at face value it appears possible that Timoleon could very well have been an early eyewitness of ball lightning. The light seemed to hover above the ship. ball lightning is known to hover or travel really slowly, but generally its life span is very short. The ball of fire that I witnessed in New York back in 1998 may have lasted a minute at the most. There is another possible explanation for Timoleon's fireball, and that is the seafaring phenomenon known as St. Elmo's Fire. This is a topic that I will discuss in depth in chapter 13.

Illustration 12

Another possible early encounter with ball lightning occurs in a book that may or may not be based on truth. In the *Queste del Saint Graal* believed to have been written in the late 12th or early 13th century by an author who is unknown, an incident takes place that appears to be related to ball lightning. The "Graal" is an allegory of sorts that centers around the semi-mythical King Arthur and his knights of the round table. It relates the adventures, trials and tribulations of the knights in their search for the legendary holy grail. The grail is a relic usually thought to be the chalice or dish used by Jesus and his disciples at the Last Supper. In recent years there have been other elaborate theories regarding the true nature of the grail. This is a topic that I would love to expound on in more depth but it does not fit in with the scope of this work. What interests us here is an event which is related early in the text when King Arthur and his knights are seated around the round table getting ready for a feast after a tournament in which the youthful Sir Galahad had proved his mettle to his fellow knights. After the knights had entered the banquet hall and were seated, King Arthur ordered that their places be set. At about this time a loud clap of thunder was heard and, the hall was suddenly lit by a bright light which was said to be seven times brighter than the normal daylight entering the hall. It was then said that the Holy Grail appeared "covered with a cloth of white samite." It hovered over the table, much to the astonishment of everyone present and served each knight in turn with the victuals that he desired. After each man had been served, Arthur rose from his chair and thanked the Lord for having fed them with his grace at the high feast of Pentecost.

The *Queste del Saint Graal* is most often thought to be a work of fiction. The noble characters within the book: Lancelot, Perceval, Bors, Galahad, and of course, King Arthur himself have been relegated to the status of myth. Perhaps the characters are fictional, but then again, perhaps they are metaphors for the attributes and deficiencies that we as human beings possess. Lancelot, for instance, is a brave knight whose physical prowess, loyalty, and chivalric qualities are unmatched. However, he is wrought by inner demons. These demons, which are nothing more than the inherent weaknesses shared by all include lust and greed. They eventually consume his being until he is rendered impotent by them.

Lancelot, therefore, could be a metaphor for the dual nature of strength and weakness that we as humans possess. Galahad may be a metaphor for youth and the continuance of the species. Arthur, perhaps, is one for wisdom and age. Whatever is the true meaning of the *Queste del Saint Graal* there appears to be some sort of truth as to the actual existence of the characters involved in the story. A historical King Arthur is thought to have existed in the late 5th century A.D. The historian Nennius who lived in the 9th century wrote of an Arthur who was a Celtic-British king who defeated an invading Anglo-Saxon army at the Battle of Mons Badon in the late 5th century. If Arthur was an actual king, than perhaps the story which is told of him and his court has some merit of truth to it. If we take this at face value, then we might be able to interpret some elements of the legend into physiological properties. For instance, in a later part of the book the noble Sir Bors is forced to fight his brother, Lionel, after Lionel accuses him of cowardice. During their engagement, Bors hears a voice from heaven telling him to quit his arms. Suddenly, a ball of fire comes down from the sky and strikes directly in between the two combatants burning their shields and rendering them both unconscious. The author of the "Graal" is obviously using a little bit of literary license here. It is doubtful that Bors heard a voice, unless it came from his own head. It is also doubtful that this lightning bolt fell directly between them as the author states. What may be factual about this story is that there probably was some sort of combat between the two knights, and that the battle might have taken place during a lightning storm. Perhaps it was even bad enough for the two noble combatants to mutually agree to a truce until the storm abated. So, from this possible interpretation of an event, we can now go back to the knights seated around the round table preparing for dinner. For just a minute, put yourself in the shoes of Sir Lancelot. You are seated at a table waiting to be served dinner. Outside, it is raining and there can be heard the rumbling of thunder in the distance. Maybe the thunder gets louder as the storm approaches, and an occasional flash of lightning can be seen from the windows in the hall. Suddenly there is a loud clap, and you see a pale, white ball moving across the hall with an irregular motion as if it might be of intelligent design and controlled by something present, but at the same time remote. What would you think?

Since the reader of this book is at least acquainted with the concept of ball lightning, the notion that this pale, white light could possibly be related to it might cross your mind, just as it crossed mine in New York back in 1998. No doubt, you would still probably be amazed and transfixed by the spectacle. The knights of King Arthur's court, and the writer of the *Queste del Saint Graal,* had no concept of ball lightning. They were unaware of its existence, so they would have naturally affixed some kind of divine meaning to this light. Since they were all seated and waiting to be served their supper when the light appeared, it can be inferred that the ball of light could have taken on the connotation of the holy cup in the same way that the Eucharist metaphorically represents the body and blood of Jesus Christ. It is the symbolism which is important here. The "pale white light" acts as a device, like the holy cup at the Last Supper which keeps the court of King Arthur (like the apostles) closer to the presence of God.

Illustration 13

So, was the Holy Grail a cup? A plate? Or some other crafted material object as is traditionally thought? Is it the womb of Mary Magdalene? A notion that has become popular in books and Hollywood in recent years, or is it simply a 5th century report of ball lightning? I have thrown the possibility out there. There is no evidence one way or the other, but the ball lightning theory holds up as well as its competing theories.

Moving forward in time to the year 1285, C.E. Britton mentions a possible case of ball lightning which he found in the *Lanercrost Chronicle*. According to the report, at the town of Lanercost, England, a high wind blew across one of the local rivers. In its wake came a flaming globe which crossed the river and destroyed two houses. High winds are usually associated with violent thunderstorms, so this seems like it is a legitimate case of ball lightning.

Another case mentioned by Britton is taken from the *Annals of Ulster* in Ireland. This event allegedly occurred on Christmas Day in the year 991 A.D. The Annal reported that a wonderful sight was seen in the sky resembling that of a burning hand.

One of the earliest proponents and supporters of ball lightning was the French astronomer, Camille Flammarion, who lived in the late 19th and early 20th centuries. Flammarion, at an early age, showed signs that he was unconventional in the way that he went about analyzing the wonders of the universe. He was a supporter of the mid-19th century spiritualist, Alan Kardec, and in fact, gave the eulogy at that man's funeral in 1869. Flammarion wrote a number of books on a variety of subjects. In *Thunder and Lightning* he devoted a whole chapter on the elusive, and at that time (1905), unsupported phenomenon of ball lightning. In describing ball lightning, which he simply called fireballs, Flammarion writes:

> In shape they are not always quite spherical, though this is their normal appearance; and although their contours are usually clearly defined, they are sometimes encircled by a kind of luminous vapor, such as we often see encircling the moon. Sometimes they are furnished with a red flame like a fuse that

has been lit. Sometimes their course is simply that of a falling star. Sometimes they leave behind them a luminous trail which remains visible long after they themselves have disappeared. They have been described as looking like a crouching kitten, an iron bar, a large orange-so harmless apparently, that you were tempted to put out your hand to catch it. There is record of one being seen as large as a millstone. One remarkable thing about them is the slowness with which they move, and which sometimes enables their course to be watched for several minutes.

It is interesting to see ball lightning described as being "so harmless apparently, that you were tempted to put out your hand to catch it." Nothing can be further from the truth, as even Flammarion himself knew from some of the reports that he had gathered. He mentions a horrible account of a fireball which entered the great hall at Feltri (Marche Trevisane) on July 27, 1789. It occurred at around 3:00 P.M. when nearly 600 people were seated in the hall. The fireball was described as being approximately the size of a cannon-ball. Ten people were killed and seventy others were injured. Twenty years later, on July 11, 1809 a fireball entered the church at Chateauneuf-les-Monstiers in the French Alps. It occurred just at the time when the bell was ringing and people were taking their seats. Nine people were killed by the fireball and eighty-two injured when it exploded among the mass. Interestingly, Flammarion also states that all the dogs that entered the church that day had been killed.

Another case that unfortunately resulted in tragedy occurred near Montfort-L' Amaury in France in 1890. A farmer was working in his fields when a thunderstorm came upon him. In an attempt to shield himself from the rain he moved up against his horses. He then observed a ball of fire pass by one of his horses ears and explode with a loud bang. The farmer and one of the horses were killed instantly by the blast. Obviously, there must have been an eyewitness to this account, or the farmers movements and means of death would not have been known.

A sad account related by Flammarion took place at Ouralsk on May 22, 1901. At around five that evening the town was hit with a severe thunderstorm which took most people by surprise. Many people crammed into doorways or other improvised shelters to get away from the storm. A young girl, seventeen years old, sat down on the threshold of one doorway along with a few more people. Witnesses then say that there was a loud clap of thunder which was followed by a "dazzling brilliant ball of fire" which made its way through across the threshold and struck the young girl on the head. The rest of the people then watched this ball of fire enter through a side door where the owner of the house was seated. After brushing up against his boots the fireball crashed through a wall and destroyed a stove-pipe which was hit with such force that it was sent careening through a window into the yard. The fireball then disappeared. Unfortunately the young girl who had been struck on the head died from her injuries. After examining the body it appeared that the fireball had actually struck her on the neck and traveled down her back and hip. It also left a black streak on her back, and knocked one of her shoes off. Strangely, Flammarion also added that all the witnesses to the girl's death had gone deaf.

One of the strangest accounts mentioned by Flammarion took place during the afternoon of September 2, 1716. Two men were descending on packhorses from the summit of a mountain when they encountered a lightning storm. They had reached an area on the mountain that was enveloped in a mist and could see flashes of lightning which seemed to be traveling in all directions. Some of these flashes took the shape of fireballs, and the men were astonished to find them flying all around them. The two men described them to be of a reddish color, of various sizes, and spinning. The two travelers were at first more curious than anything else at the spectacle. However, their curiosity soon turned to fear when one of these fireballs which they estimated to be about two feet in diameter exploded very close to them causing streams of light to fly off in all directions. The two men were now scared out of their wits. After a short while the storm passed over them, and they were then left to reflect on the experience.

Taking our leave of Flammarion and the continent of Europe we now travel to the other side of the Atlantic where the Reverend Increase

Mather recorded an incident which took place in the town of Marshfield, Massachusetts. On July 31, 1658 when Marshfield still lay within the boundaries of Plymouth colony, a man and a boy were working in the fields when they were suddenly forced to take shelter in a house after a thunderstorm had snuck up on them. The man, whose name was John Phillips, sat down near the chimney. Shortly after doing so a small black cloud appeared, and a fireball traveled down the chimney into the house killing him. The boy, and another man who was in the house at the time, were standing only six feet away from Phillips when the fireball struck. They escaped without a scratch. After a careful inspection of the house it was noticed that some of the bricks appeared to have been beaten out of the fireplace, and some rafters had split.

One interesting aspect of ball lightning is a characteristic that is not commonly mentioned. The reason for this is that most of the witnesses who have experienced it first hand and up close and personal have not lived to tell about it. This characteristic would be the sensation of heat that is associated with it. In September of 1980 a group of people were traveling in a car from Columbia, South Carolina to Charleston to attend a wedding. About 50 miles south of Columbia on Interstate 26, a young man named Anthony Bailey, who was sitting in the back seat of the car, was startled to see a pale, white fireball only inches outside the window. The fireball was about twice the size of a basketball. Bailey estimates that he only observed the fireball for a second or two before it exploded with a loud bang. This makes sense when estimating that the car was traveling about 60 to 70 miles per hour. It seems apparent that the fireball happened to descend or ascend in front of the window just as the car passed by. Bailey told the author that the strangest thing about the whole incident was the heat that was generated from the fireball even with the window up.

We have now seen the dangerous effects that these fireballs have when they encounter humans. Armed with this information we can now look at ball lightning as a possible answer to some of the world's biggest mysteries. Since time immemorial people have reported seeing ghosts, phantoms, poltergeists, and other supernatural entities. Poltergeists are often thought to be destructive spirits who have not passed over from the corporeal world. They are supposed to be responsible for wreaking havoc and destroying property, and in some cases even injuring and killing people. They are also sometimes described as looking like a ball of light. Is it possible that some of these cases that are attributed to evil and malignant spirits are no more than cases of the natural phenomena of ball lightning?

Another unsolved mystery has to do with a rare destructive phenomenon known as spontaneous human combustion. In certain cases it has been recorded that human beings have suddenly burst into flames for no apparent reason. In most cases it seems that only the human body has burned and not the surrounding area. This would not be the case, say, if someone fell asleep in a chair smoking a cigarette. In this instance there would be evidence of some kind. At the least the area surrounding the charred corpse would be scorched. In cases of human combustion only the human is affected by the fire. Sometimes it is so catastrophic that the unfortunate victim is reduced to ashes.

A rare case of this phenomenon was told to the author many years ago by a first hand witness who had seen the after affects. In April of 1987 I was stationed at Marine Corps base Camp Hanson on Okinawa, Japan. I was pulling a stint on mess duty, and ended up working as a dishwasher in a room which we marines call "the pot shack." The hours were long and tedious. We were up at 4:30 A.M. and worked sometimes up until 9:00 or 10:00 P.M in the evening, depending on when we finished cleaning the endless procession of cumbersome pots, dishes, and silverware which continuously streamed in during the course of the day. Besides our work we pilfered the time away with idle gossip like "did you hear that Gunnery Sergeant Pyle was messing around with Captain Doolittle's wife?" or nonsense like "the first thing that I am going to do when I get back to the states is get drunk and look up the harem of women that are attached to me." It all depended on the company that was assigned to assist with this laborious job that decided on the quality of conversation that ensued.

One day a certain lance corporal, whose name I have long since forgotten, was assigned to work next to me. Like the others I worked with earlier, we soon were engaged in conversation. Eventually, the two of us realized that we could talk on a certain level about things that most people in that environment would find different. The topic of conversation dealt with astronomy, but somehow the conversation drifted to the fringe elements of science, and we began to talk about UFO's, cryptozoology and other weird topics. Eventually the fellow brought up the subject of spontaneous human combustion. If I recall correctly, he told me that he had been a volunteer fireman in a small town somewhere out west. One day, he said that his services were needed at a house where a fire had been reported. Upon arrival, he found much to his horror, a charred body lying down on the floor of the house. According to him the body had been burned beyond recognition and its posture was set in a position which suggested an agonizing death. He also told the author that the strangest thing about the whole thing was that there were no signs of fire anywhere else in the house. Not even around the corpse.

Normally I would hear something like this and brush it off as fantasy or a simple misinterpretation of the facts. Could the body, perhaps, have been burned someplace else and then brought to the house? Not likely, since he was called to the house after a fire had been reported. According to him there was no evidence to suggest that the fire was caused by any external device like a match or a candle. So, how then did the fire start? His answer was that he believed this person was a victim of spontaneous human combustion. I must say that this fellow seemed extremely sincere and did not seem like the kind of person who would make up such a thing.

Before this conversation in "the pot shack" back in 1987, I had never before heard of this spontaneous human combustion. When this fellow suggested to me that this might be a case, I was fascinated and have kept the details of this incident in my memory for the better part of two decades. What causes spontaneous human combustion? A proven answer does not exist and the theories that abound are problematic. Is it possible that it could be caused by ball lightning? Since both phenomena are rare, especially human combustion, this possibility might be worth looking into. Let us recall Flammarion's account of the young girl killed while standing in the doorway in France back in 1901. Her body had been marked with black streaks.

In order to prove this theory at least three tasks need to be done. First, and foremost, there must be a scientific study done to examine the ball lightning phenomenon in depth, and from this study, proof must emerge relating to its true nature. Secondly, in cases where spontaneous human combustion are thought to have occurred, a chemical analysis of the remains must be carried out which could possibly tell us if any type of electrical energy was responsible for the destruction. It would also help to find out if any thunderstorms were present at the time of the incident. Finally, the third thing that needs to be done is to find a qualified person or persons willing to correlate the two phenomena and analyze the data accrued from the unfortunate victim with the properties known to cause ball lightning. Unfortunately, this may be a tricky thing to do. The reason I say this is that spontaneous human combustion is one of those theories that most people believe falls into a dark corner study along with bigfoot and UFO's. A professional scientist with a reputation to uphold is apt to be derided or looked at with a degree of suspicion.

Leaving the mysteries of ball lightning we now turn to a type of terrestrial fireball that is just as enigmatic as ball lightning but quietly lurks in the shadowy, marshy grounds of swamps and ponds. I introduce you now to the "ignis fatuus" or his better known alias "will-o'-the wisp."

Chapter 12

THE WILL—O'-THE WISP

Out of all of the different types of fireballs addressed in this book none of them is as eerie and mysterious as the nocturnal light known as the will-o'-the wisp. For centuries "Will" has been seen haunting swamps and ponds and lurking in the tall grass of a moor. A 19th century poet named Cecil Cavendish brings us a romantic glimpse of this spectral light when he writes:

> "A marshy meadow-a quiet pond-
> A lonely road-and a hill beyond-
> In the reedy marsh below the hill,
> On starlight nights when the air is still,
> Where rushes and cresses grow green and crisp,
> There goes, dancing, will-o'-the wisp,
>   will-o'-the wisp so gay
>
> We see the gleam of his lantern bright
> Flitting about in the quiet night.
> He balances on the cattail tops,
> Then to the rustling reeds he drops,
> And reeds to the rushes will softly lisp,
> Here comes, dancing, will-o'-the wisp,
>   will-o'-the wisp so gay

The east grows gray at the touch of dawn.
Presto! will-o'-the wisp is gone.
For the morning wind blows out his light-
He'll dance again on another night.
When crickets are chirping in grasses crisp,
Then we'll watch for will-o'-the wisp,
      will-o'-the wisp so gay

So . . . who is will-o'-the wisp? Or more properly put: what is will-o'-the wisp? The modern day traveler may not have ever heard of "Will", but that was not the case for people who lived in a more antiquated age. will-o'-the wisp is more properly called the *"ignis fatuus"* which translated from the Latin means "the luring fire" or "the wild fire." It has often been described by witnesses as appearing like a torch in the unsteady hand of an unseen traveler. Sometimes it is a ball of light or a fireball that lingers over an area of marshy ground, or a flame which travels slowly across a lonely meadow. Most sightings occur in places that are damp, or areas where there is a lot of vegetation that has suffered from years of neglect. The color of the *ignis fatuus* varies, but it has been known to appear in various shades of red, orange, white, blue, or green. A common characteristic associated with this phenomenon is its elusive nature. A person seeing an *ignis fatuus* may attempt to follow it, but much to the traveler's dismay, will never catch up to it. Indeed, this aspect of its nature is mentioned in its name. In years gone by the reports were numerous of tired, weary travelers who lose themselves on a dark lonely stretch of highway. Suddenly, they see a distant light and believe that it is a fellow traveler who is lighting his way with a torch or lantern. The unsuspecting traveler follows, thinking that this torch might lead the way to the comfort of a warm fire in a roadside inn. The gullible traveler may even follow when the torch veers off the road into a meadow. The traveler simply thinks that the torch bearer knows a shortcut. After a while the by now exhausted traveler finds that the torch leads into a foggy bog where it disappears. It is at this time that the traveler realizes that he has been duped, and has merely become the latest victim of the chicanery of will-o'-the wisp.

The *ignis fatuus* or will-o'-the wisp is known by many other names around the world. In Ireland, England, Scotland and many of the Scandinavian countries it is sometimes referred to as "Jack-with-a-Lantern." Of course, many people reading this are well aware of the modern connotation of this name "Jack-o'-Lantern." Every year, at Halloween, houses are decorated with these carved out pumpkin heads, but many people are unaware of this tradition's historical origin. In Japan these mysterious balls of fire are known as "Hitodama" which translated into English means "human soul." They are believed to be the lost souls of the dead searching for a passageway into the spirit world.

There are many legends attached to these strange lights. The name will-o'-the wisp probably derived somewhere in the moors of southern England. The dating of the name is not known for sure, but references to it can be found in works dating back to at least the 17th century. The name itself can be broken down and studied in order to get a better understanding of its meaning. "Will," of course, is short for the name "William" and is probably used the most because it rhymes with the word "Wisp." It is more fluid and poetic to say "will-o'-the wisp" than say "steve-o'-the wisp." A wisp, used in this way is a small bundle of straw that is bound tightly and when lit is used as a sort of torch. Hence, "Will" or "William" with a "Wisp" or "Torch." Before I attempt to discern the actual cause of this phenomenon, I will relate a few of the many tales and legends that surround it.

A popular origin for will-o'-the wisp was published in a book of old Hibernian tales in the 19th century. It was reprinted in a book on Irish literature printed by John D. Morris out of Philadelphia in 1904. This old Celtic tale tells the story of Will Cooper, a humble blacksmith, who, like the typical Irish stereotype, has a problem with "John Barleycorn." He is a hard working fellow, but the money that he makes is mostly spent at the local inn where he tips the bottle until the late hours of the night, and along with his rummy friends toasts the heroic deeds of the Irish martyrs Robert Emmet and Wolfe Tone.

One day Will is working at his forge when an old, haggard looking man walks up to him seeking succor from the elements. Will takes pity on the bedraggled man and takes him in. After a hearty meal the old man is so taken with Will's hospitality that he sees fit to grant him three wishes. Will, probably believing that the old man is playing some kind of game, humors him and asks that his first wish be granted that "no man who takes his sledge into his hand may be free of it until he sees fit." For his second wish he says that " if anyone thinks fit to set down in his arm chair they will not be able to rise until he releases him." His third and final wish is that "whatever money that he puts into his purse no person but himself has the power to remove it." The old man frowns upon Will's three desires, and lets him know that for at least one of them, he should have asked for heaven. However, before taking his leave he grants the blacksmith his three wishes.

As time passes Will forgets about the three wishes, and one day while he is working at his forge, the Devil in the disguise of a man approaches him and proposes to give Will seven years of anything that he desires if at the end of that period he relinquishes his soul. Will thinks this proposal over, but since times are so bad, and his financial condition so destitute, he does not ponder over it for very long before assenting to the Devil's offer. So, for seven years Will lives the life of a gentleman and becomes a rich man. Everything that he touches turns to gold. He even founds his own town and changes his name to Bill Money. To keep up appearances, Will still occasionally plies at his trade. He does this so that no one will find out that he has made a deal with Satan. However, although he keeps up this deceptive front, there are some who suspect he has made some kind of sinister deal because of the sudden change in his lifestyle.

When seven years has elapsed, the Devil came looking for Will and found him in his shop working on a project that he had promised to do for a friend. The Devil asks him if he is ready to go, but Will attempts to renege on the bargain. He pleads for the Devil to at least give him enough time to finish this project he has started. Reluctantly, the Devil agrees, and Will even coaxes the fallen angel to assist him with it. Remembering the three wishes from long ago he asks the Devil to pick up his sledge

most certainly revert to his old tricks. The Devil could simply not have this. Will pleaded with his former antagonist and caused such a ruckus outside of Hell's gates that the Devil agreed to parley with him. The Devil pondered on what to do with Will when it suddenly dawned on him that the cunning and deceitful blacksmith could be put to good use and serve as a useful ally. He therefore had Will removed from outside of Hell's gates and conducted back to Ireland . Once there, he gave this malignant spirit a lighted wisp and informed him that for eternity he should wander the moors and swamps and deceive the lost and wearied traveler. Forever after, he shall be known as will-o'-the wisp.

Another morbid and somewhat disturbing legend comes from Dutch tradition. It is said that the will-o'-the wisp's are the souls of children who have not been baptized. They naturally like to play around marshy watery areas in the hope that some man of the cloth might take notice and baptize them. Indeed, there is a story related by T.F. Thiselton Dyer and published in *Gentleman's Magazine* which tells the tale of a Dutch minister who is traveling back to his home late one night when he stumbles across three of these strange lights. He is aware of the legend that surrounds them, so he takes it upon himself to baptize the three wisps. However, upon saying the baptismal words he is astonished to see thousands of wisps appear before him in desperation to be baptized. Terrified, he runs as fast as he can for the safety of his home.

Illustration 14

In parts of Germany the *ignis fatuus* goes by the name of "Heerwisch." According to legend, whenever Heerwisch hears a certain song that he doesn't like emanating from the lips of a traveler, he gets angry and takes the form of a goblin. It is said that he leads his victims to marshy ground where he terrorizes and sometimes kills them. One particularly frightening tale tells of a young girl traveling home one night in the village of Godorf. As she crosses a field to her house, she begins to sing a lullaby which irritates Heerwisch and provokes his wrath. Taking the shape of a goblin he chases the frightened girl back to her house. However, before she can shut the door, he sneaks in behind her and terrifies the rest of her startled family.

The will-o'-the wisp is sometimes called an "elf candle," and is therefore frequently associated with the fairy phenomenon which reached the height of its popularity during the first two decades of the 20th century. With supporters like Sir Arthur Conan Doyle, the author of *Sherlock Holmes*, the *ignis fatuus* were alleged to be the vehicles which the fairies used to travel from their land to ours. One early writer described these fairies as being "very small, only a few inches high with an airy, almost transparent body." He further added that "they do not live alone, or in pairs, but always in large societies." Irish legend also says that these fairies or elves come out at night from their secret abodes and congregate around marshes, streams, and cemeteries, or other places that are not particularly favorable for human habitation. They disappear when the sun rises leaving for a place known as "Thiera na Oge," which is Gaelic and is pronounced "Cheer na Nog." Translated into English it means "The land of youth." Supposedly no one who enters this realm ever grows old.

In parts of Ireland and Scotland the *ignis fatuus* is known by the rather morbid term of "corpse candle." It is said that if one of these "corpse candles" ventures near a house, it is a sign that one of its occupants will soon die. The Irish fairy known as the "banshee" is also supposedly tied to the "corpse candle" legend. The "banshee" is a female fairy who hangs around a home when there is someone that is sick. It is said that this strange creature howls or screams into the night as its way of grieving for the sick. It is easy to see how this legend came about. It seems probable that if the *ignis fatuus* is accompanied by a strong wind, one could see the makings of the Irish "banshee."

Leaving the realm of legend and folklore we now turn to some actual reports of these strange terrestrial fireballs which have been handed down to us over the last few centuries. In Hanover Germany in 1807 an astronomer named Bessel observed a will-o'-the wisp while traveling across a moor one winter night. He explained that the night was calm when he observed a great many bluish flames above an area of uneven ground. Most of the flames were stationary, but some seemed to dance around from place to place. However, Bessel was quick to notice that these strange balls of fire were mostly seen over areas where rainwater had accumulated in puddles.

Another 19th century German account is given by a Major Blesson of Berlin. One day, he happened to be out on the marsh and noticed that there were certain places on the ground where bubbles of gas could be seen escaping from the Earth. That night he also observed a pale, blue flame hovering over the same ground. He quickly deduced that the two phenomena were related, so he ventured forth to have a look. As he neared the spot where the bluish flames appeared, he was rather astonished to see them recede. The next evening the Major returned to the swamp and decided that this time he would approach the area with caution reasoning that the reason that the flames receded from him had to do with the rapid movement of his body. He was correct in this assumption. He carefully approached one of them and was able to get right up next to it. He then decided that he would try a little experiment by holding a piece of paper within the flame. The flame ignited the paper, and the Major then concluded that the flames were caused by gas in the marsh.

The two German reports that I have just mentioned were related by Professor Edwin J. Houston in his book *The Wonder Book of the Atmosphere.* Houston's theory behind the will-o'-the wisp phenomenon was that it is similar to a gas known as phosphoretted hydrogen which can ignite merely by coming into contact with air. He believed that it was due to marsh gas which is caused by decomposing animal and plant matter. This theory holds weight, as most reports of the "Ignis Fatuus" occur around places where animal and plant matter is abundant in great quantities.

Another intriguing and somewhat interesting theory was related by T.L. Phipson in an article for *Belgravia Magazine* in 1868. Phipson states that two entomologists named Kirby and Spence attempted to explain the phenomena as being caused by "luminous insects hovering in clusters over marshy ground." In many respects this seems like a plausible theory, for fireflies are known to breed in swamps and marshy ground where the will-o'-the wisp is most commonly seen. The reason for this is that these wet areas provide a source of nourishment for the firefly larvae. The "firefly," or as they are sometimes called lightning bug theory also holds up to the elusive nature of the will-o'-the wisp. How many of you reading this right now can recall attempting to catch a firefly in your youth? I remember engaging in this pastime, and can also remember how tricky it was to catch one of those little buggers. I did, however, become fairly adept at it by cupping my hands in a certain way, and like Major Blesson, by furtively approaching the nocturnal insects.

There are also other insects known as glowworms that are bioluminescent and could also be associated with the will-o'-the wisp phenomenon in areas where there are rainforests or caves. The glowworm is the larvae of either a fly or beetle which hangs in threads from dense jungle canopies or cave ceilings. The author of this book had the chance to observe these glowworms firsthand while on a visit to the Waitomo caves on the north island of New Zealand in August of 2002. Observing these larvae from a boat inside of a dark cave gave me the illusion of looking up at the night sky on a clear night full of stars. These glowworms hang down from the cave ceiling in threads of a silky type of mucus. Although there is no wind in a cave, these thin threads in a dark jungle environment might catch a breeze and therefore appear to move.

The firefly/glowworm theory, however, has holes in it. The most obvious is that it does not explain the will-o'-the wisp which frequently seems to emerge from the earth itself in places other than marshy ground.

The French scientist W. De Fonvielle records an incident that took place on July 2, 1750. A French priest named Abbe Richard observed a large flame rising from the pavement near his church. The flame rose about 12 to 15 feet in the air. Another priest the Abbe Girolamo Leeoni de Ceneda observed a similar flame of bluish light rise from the earth. This flame seemed to hover in the air for a short while before exploding with a terrific noise.

Although the will-o'-the wisp hardly gets any attention today, the elusive light still grabs the headlines every now and then. In March of 1966 a celebrated UFO case in Ann Arbor, Michigan was attributed as possibly being related to the will-o'-the wisp. The suggestion that objects seen flying over a swampy area near Ann Arbor were cases of swamp gas or will-o'-the wisp was offered by the famous UFO investigator and astronomer J. Allen Hynek. The public, however, was outraged by this explanation as some of the reports were obviously not related to swamp gas. One report noted a "falling star-a brightly lighted object that was plunging swiftly and silently to earth." Hynek was forced to restate his opinion adding that the swamp gas theory was just one possible explanation for the sightings.

For my own part, I think that it is plausible that at least some of the reports which were seen over the swamp near Ann Arbor could have been related to old "Will." As far as the other reports that dealt with objects seen descending from the sky, I cannot say for sure, but they very well could have been legitimate cases of UFO's. Also, like similar sighting outbreaks, some of the reports could also have been attributed to some form of mass hysteria. I would expound more in detail on the incident if it fit in with the scope of this work, but it does not. For a detailed and interesting account of the Ann Arbor lights the reader can refer to Frank Edward's 1967 book *Flying Saucers Here and Now.* This book, along with his other book on the UFO phenomenon *Flying Saucers Serious Business* were popular best-sellers in the late 1960's and can be found on various places on the internet, or at used book stores for a reasonable price.

The will-o'-the wisp may also be partly responsible for the well known lights seen over Brown Mountain, which is located in the Blue Ridge mountains near Asheville, North Carolina. For well over a century people have seen strange fireball or orb-like lights rising from the mountain when observing it from a distance. Over the years these lights have been attributed to everything from train headlights to flying saucers. However, there is probably a simpler more natural reason for their existence, and one of the more likely suspects is the *ignis fatuus*. If this is the case, then it would be the *ignis fatuus* on a grander scale, and a phenomena of a type that is not well understood at this date, for generally speaking the will-o'-the wisp is seen hovering not far from the ground.

The will-o'-the wisp is still almost as mysterious today as it was two or three centuries ago. However, the mystique and legends that surround it which have spread orally from generation to generation are slowly being forgotten. These tales are dying off as the older generations who grew up in the age before television and computers are dying. Urban sprawl and the gradual encroachment of marshlands by developers may account in a reduction of sightings. However, the biggest blow to the will-o'-the wisp legends is the current mode of travel. People today are more reliant on their automobiles than ever before. Even the local grocery store two blocks down the street is too far to walk for most people. We have become an enclosed climate controlled population that has become dependent on the modern conveniences that technology has given us. In the old days it was nothing for a person to walk to a neighboring town on business. If a person did take a horse or a buggy it was under the open air where one could get a sense of the elements and be more in tune with the natural habitat around him. A strange noise in the night might invoke the individuals sense of awareness and conjure up all kinds of possibilities as to its origin or nature. William Bradford, the second Governor of Plymouth Colony writes of an event that took place during the pilgrims first winter . During the first cruel winter in New England, Bradford writes that two men, John Goodman and Peter Browne, were out cutting thatch in the forest when one of their hounds took off into the woods in pursuit of something. The two men followed in hot pursuit, but being unfamiliar with the terrain and landscape soon lost themselves in the strange land. They were forced to spend a cold, rainy, and miserable night in the forest. They climbed a tree and took refuge there believing that it was safer. At one point during the night they were haunted by the eerie sound of something that they could not identify. They attributed it to the call of a demon lion. It is obvious here that the two men's imaginations were running wild. After all, put yourself in their position. You are in a strange land in a vast wilderness. What would you think?

Today's lost travelers, unless in some far off remote part of the world, would have no trouble finding civilization. If they heard something similar to what these two signers of the Mayflower Compact experienced, they would more than likely brush it off as something commonly attached to the material world in which they live.

So it is with will-o'-the wisp, perhaps seen, but no longer recognized, and today largely ignored. However, if by chance one day you decide to take a walk one night down a dark and lonely road, don't be surprised if you see a torch beckoning you to follow along a path across a field. A shortcut? Maybe? But more than likely it is old Will trying to live up to his part of the bargain made with the Devil in an age long ago.

Chapter 13

THE FIRE OF ST. ELMO

In the previous chapter we discussed the phantom light or fireball known commonly as will-o'-the wisp. That phenomenon is confined only to areas of the Earth which are covered with land. This is not the case with our next study which occurs mostly, but not always at sea. It is a phenomenon that has been recorded by sailors and sea faring travelers since ancient times. It goes by many names, but is best known as St. Elmo's fire. It is said to be named after a Dominican priest named Peter Gonzales who lived in the 13th century and went by the name of St. Elmo. He is known as the patron saint of sailors.

The fire of St. Elmo usually appears as a small flame or a ball of fire. It is often seen on the masts of ships or the tips of metallic objects like swords or lances. It varies in color depending on the eyewitness, but it is often described as a pale blue or green color although it is sometimes yellow, orange, or white.

One of the earliest records of it is in Greek mythology. It is mentioned in the tale of Jason and the Argonauts in their quest to find the golden fleece. According to the myth, the Argo, which was Jason's ship, was being battered by a terrible tempest when suddenly two fireballs appeared above two of Jason's greatest warriors, two twins named Castor and Pollux. The storm then abated, and legend has it that from this point on every time the fires of Castor and Pollux appear on the masts of a ship, it will be a successful voyage.

The fires of Castor and Pollux is merely another name for St. Elmo's fire . It was the Portuguese and Spanish sailors during the age of exploration that coined the phrase "Corpo Santo" or "The Holy Sanctified Body" believing that the spirit of St. Elmo was with them to guide the vessel safely on its course. In the fall of 1519 Ferdinand Magellan set sail from the Canary Islands on a southerly course hugging the coast of West Africa. It was still near the beginning of the voyage which would eventually bring one of his ships around the globe. After a few weeks at sea, his fleet of five vessels was caught in a series of bad storms. At one point there was observed several globes of fire on the yardarms of the ship *Trinidad*. The fires of St. Elmo were visible for about two and a half hours which reassured the crew that they were being guided by a divine light.

The fire of St. Elmo was also observed by the French Admiral, Claude Forbin, who is best known for his nautical exploits during the War of Austrian Succession. It was his bold attempt to invade England by landing the Stuart heir presumptive Charles Edward Stuart (The Old Pretender) in Scotland in 1708 which gained him a bit of notoriety. However, the reason I mention admiral Forbin is not for his militaristic abilities, but for something which gives cause for his mention in this book. In 1696, Forbin and his fleet were sailing near the Balearic Islands when they were suddenly caught in a fierce storm. After ordering in the sails, he observed about thirty glows or fires spread out over the rigging of his ship. One of these fires was estimated to be about a foot and a half long and was situated on top of the weather vane. For some reason Forbin decided to send a sailor up the mast to remove it. As the sailor attempted to perform his duty he reported down to the admiral that the fire was making a noise similar to that of burning wet powder. After removing the weather vane the fire merely moved to the top of the mast where it stayed for a short time until it finally disappeared.

Illustration 15

Apparently, Forbin was either unfamiliar with the phenomenon, perhaps believing that the fire might burn his ship, or he sent the sailor up the mast out of scientific curiosity in order to get a closer look at the fire. Although not much was gleaned by this experiment, Forbin did learn that the fire was not only attracted to metal, but was also attracted to wood. This was perhaps the first experiment ever performed in regard to St. Elmo's fire.

One report of the fire of St. Elmo that is not well known is told by Captain Edward Haies of the *Golden Hinde,* one of the vessels assigned to Sir Humphrey Gilbert's American expedition in 1583. The expeditions mission had a dual purpose, which was to find a suitable place to start an English colony, and if possible, find the elusive Northwest Passage.

After an eventful voyage, Gilbert's fleet was on its way back to England when the vessels encountered some stormy seas. At this time, Captain Haies believed that they were sailing at about the 44th parallel just north of the Azores. Haies writes:

> By that time we had brought the islands of Acores (Azores) south of us, yet wee then keeping much to the north, untill we had got into the height and elevation of England: we met with very foule weather, and terrible seas, breaking short and high pyramid wise. The reason whereof seemed to proceede either of hilly grounds high and low within the sea, as we see hilles and dales upon the land, upon which the seas doe mount and fall:or else the cause proceeded of diversitie of winds, shifting often in sundry points:al which having power to move the great ocean, which againe is not presently setled, so many seas do encounter together, as there had bene diversitie of windes. Howsoever it commeth to passe, men which all their lifetime had occupied the sea, never saw more outragious seas. We had also upon our maine yard, an apparition of a little fire by night , which seamen doe call Castor and Pollux. But we had onely one, which they take an evill signe of more tempest: the same is usuall in stormes.

This journal entry tells us a lot in a small package. First, it tells us that St. Elmo's fire is associated at least in some way with a storm, and although Captain Haies does not specifically mention lightning, it can be implied when he mentions "very foule weather." Second, it tells us what sailors had thought of the fire of St. Elmo over four centuries ago. We have already mentioned the sailors of Magellan's voyage that observed several fires on the *Trinidad* and interpreted it as a sign of good luck. In this case, however, Captain Haies makes it plain that there is only one fire spotted and that this is taken for "an evill signe of more tempest." Whether or not there is some kind of science to this theory is debatable. One thing the sailors of yesteryear certainly believed is that either good or evil would present itself in some manifestation after observing the fire of St. Elmo. Since Captain Haies observes only one fire he naturally concludes that some evil or misfortune will come from it. In this instance the Captain proved that he possessed the salt of a sailor. For some reason, Sir Humphrey Gilbert had decided to travel on board the smaller frigate *Squirrel* instead of the *Golden Hinde*. That evening the storm worsened. The frigate carrying the admiral had approached to within hailing distance of the *Golden Hinde*. Sir Humphrey could be seen sitting on the deck with a book in his hand which was either the Bible, or Sir Thomas More's *Utopia*. He hailed the crew, and according to Captain Haies his booming voice could be heard above the raging swells and howling winds exclaiming "WE ARE AS NEERE TO HEAVEN BY SEA AS BY LAND!" The lighter frigate passed the flagship and as night came on its light could be seen bobbing up and down with the waves. It must have seemed as if Poseidon himself was holding a torch aloft to guide the *Golden Hinde*. Soon, however, the light disappeared and was not seen again. When dawn came, Captain Haies and his crew searched the horizon for some sign of the frigate. Unfortunately nothing could be seen except a seemingly endless ocean. The lone fire had been proven prophetic. Sir Humphrey Gilbert was never seen or heard of again. Captain Haies and his crew arrived in England about two weeks later to deliver the sad news.

Another account of St. Elmo's fire was recorded by a royal navy lieutenant named Milne off the coast of Brazil in September of 1827. Milne described the day as being sultry with storm clouds in the southwestern sky. By nightfall the clouds had caught up to the ship, and much to the crews dismay they were accompanied by thunder and lightning. It was at this time the Lt. Milne noticed two St. Elmo's fire's. One of them was on the vane staff at the mast head, while the other was located on the fore-topsail yard. Lt. Milne then related how a young officer climbed the mast to get a better look at the fires. This man decided that one of the fires originated from an iron bolt and was about the size of a walnut. Its color was described as yellow in the center with a touch of blue on its periphery. The young officer decided to touch the light to see what would happen. He found that it caused it to smoke and make a sort of hissing sound. However, when his wet sleeve touched it, the flame shot up his jacket and died out. Apparently, he then decided that it would be a good idea to climb down from the mast and not bother the other fire which was at the vane staff. No doubt his crewmates must have been a little perturbed by this action for it left only one fire lit!

Another experiment involving the fire of St. Elmo occurred on a ship sailing in the English Channel in March of 1866. The captain of this vessel reported seeing a fire on the extremity of every yard of the ship. The most pronounced flame appeared on the bowsprit and the captain's curiosity was such that he felt the need to examine the fire of St. Elmo a little bit closer. He therefore made his way up one of the masts to get a better look. He wanted to see what kind of affect the fire would have on his body so he reached out to it and found that the flame itself did not generate much heat. The next morning the captain had the masts examined in order to see if the fires had any effect on the paint or varnish. To his surprise he found that the fire of St. Elmo had produced no ill-effects at all to the ships mast or rigging.

Some of the early sightings of St. Elmo's fire led to much discussion on the nature of the phenomenon. What caused it? It was thought by many people to be closely associated with ball lightning or the will-o'-the wisp. Although it is possibly related to these two phenomena, which both, obviously, involve some sort of electrical discharge. It is unique in itself in that it seems to arise from the ionization of air molecules. This ionization is nothing more than the free movement of charged particles. This is why the fires are almost always seen when a storm is near, or actually present. The air and ground at this time is in a charged state. The reason that the fires are attracted to the masts and rigging of a ship is simply that these objects are the highest points on the water.

We now see that the word "fire" in its name is a misnomer. It is not a fire at all, but plasma. Our ancestors had no way of knowing this, just as they had no way of knowing that the sun wasn't a gigantic bonfire in the sky. They labeled things as they saw them, and usually that meant naming an object or a phenomena after something that they believed it was closely associated with.

Although the fire of St. Elmo is frequently associated with the sea, it is not wholly confined to that region of the Earth. It has also been known to appear on mountains, trees, buildings, and even on the rims of people's hats and clothing. It is particularly fond of church steeples, which, when appearing, is naturally assumed to be a sign of god's blessing. Anywhere there is a pointed or narrow object in the vicinity of a thunderstorm the fire of St. Elmo is likely to appear. This is especially true of metallic objects which are better conductors of electricity and hence have a better chance of a coronal discharge.

One interesting account of the fire of St. Elmo was given by a former president of Trinity College in Hartford, Connecticut. In the winter of 1839, this man, a Mr. Silas Totten was caught outside in a snowstorm. He was using an umbrella and was keen to notice every now and then flashes of light which he thought looked similar to lightning. He was startled to find that the flashes emanated from his umbrella. He was amazed further when he noticed another person walking down the street whose umbrella's tip was covered with a flame.

An example of a church steeple covered with St. Elmo's fire occurred on the evening of March 2, 1869 at the church of St. Catherine de Fierbois in Chinon, France. According to Arthur Parnell in his 1882 book *The Action of Lightning,* there was seen, shortly after a storm had ended, "a crown of fire" encompassing the cross on the church's steeple which was

130 feet in the air. Parnell also noted that there are plenty of St. Elmo's fire's present at the Notre Dame cathedral in Paris whenever there is a thunderstorm present.

As I have mentioned, there are also accounts of St. Elmo's fire attaching itself to people. Gustav Hartwig in his book *The Aerial World* states that on January 17, 1817 the east coast of the United States was hit with a wall of storms. Many people who were caught outside reported that the rims of their hats, seems of their gloves, and other objects with edges or points were covered with a "wavering flame" and produced a "crackling noise similar to that which is heard when water is about to boil." Hartwig mentions another account that occurred in Switzerland on May 3, 1821. A Dr. Allemand in a letter to a colleague in Geneva writes:

> I was overtaken while on my way to the village of Motiers near Neufchatel, at about ten o'clock at night, by one of the most dreadful thunderstorms imaginable. The night, which had already before been excessively dark, became still more so through the rain that was falling in torrents, and the frequent lightning alone enabled me to see the road. Under these circumstances I suddenly saw a light which seemed to come from above, and raising my eyes I perceived that the rim of my hat was luminous. Believing it to be a real flame, and before I had time to reflect, I quickly passed my hand over it with the intention of extinguishing it, but to my great surprise it appeared more lively than before. I shook my hand, which was dripping with the water that ran from my hat, and saw it shine like a polished metallic surface when reflecting a bright light. I now began to feel a little frightened, and deliberated for a moment whether I should not take refuge in a neighboring farm; but some knowledge of physical laws, and a most perfect confidence in the supreme author of the formidable apparatus which surrounded me, soon induced me to continue my route. Having already, without any evil consequences, filled my hand with the electrical water which shone on the rim of my hat, I now repeated the experiment (though not without some apprehension), to ascertain whether this phosphoric light had

any smell, and whether it sparkled or produced a crackling noise. I, however, remarked no other phenomenon than the beautiful light which did not rise above my hand when I opened it, but seemed to be spread over its surface like a brilliant varnish. Continuing to walk on with my attention continually fixed on the bright aureola of my hat, I now saw vivid rays proceeding from a small plate of metal on the handle of my umbrella. My first movement was again to pass my thumb over it to extinguish this new fire, but the same phenomenon occurred as before, for my hand became as luminous as the metallic plate which it rubbed. All these phenomena ceased as I approached Motiers, and I attributed their cessation to the vicinity of some large poplar trees which border the road near the village.

From the reports that have been mentioned thus far, it can be plainly seen that the fire of St. Elmo is not harmful when humans come in contact with it. To the best of my knowledge no one has ever been killed or injured by this electrical phenomenon. However, there have been reports of people experiencing a sort of uneasiness or slight pain when being exposed to it. Hartwig mentions an incident that occurred on June 22, 1867 on the summit of Piz Surley. A Swiss scientist Henri de Saussure along with a couple of members of his climbing party were taking a rest when they began to experience a stinging sensation over their bodies as if they were being stung by wasps. De Saussure realized that it was being caused by some sort of electrical current that seemed to have enveloped the summit. A distant rumble of thunder could be heard so it was assumed that the source of the problem lay in that direction. Once the party had descended about 100 meters from the summit the painful sensations subsided. It is worth noting here that De Saussure did not see the visual effects of a St. Elmo's fire.

In more recent times the fire of St. Elmo has been observed not only on ships but on airplanes, dirigibles and other machines that take to the sky. I have mentioned that there has never been anyone killed by the fire of St. Elmo. Indeed, the proof is lacking, but when it comes to air travel it has been suspect in a number of unsolved airplane crashes throughout the past century.

On June 26, 1959 a TWA Super Constellation took off from Milan, Italy bound for New York, via Paris. For the 68 passengers and crew members it was supposed to be just another routine flight. The plane took off and reported nothing unusual. However, only twelve minutes after takeoff something terrible happened. The right wing of the aircraft came off causing the plane to lose stability and crash in a fiery explosion in a wooded area outside of Milan, killing all on board. Investigators were perplexed by the crash. All takeoff procedures and radio communication with the crew showed no sign of the plane having any difficulty. The debris scattered woods offered few clues to go on. After an eleven month investigation, experts ruled out various likely culprits such as engine failure, a wiring problem, or sabotage. They then decided to look at the possibility of St. Elmo's fire as the cause. For years pilots had reported seeing the fires on their wings, propellers and other parts of their aircraft. Generally, however, the phenomenon was looked at as being quite harmless. Even lightning hitting a plane, though rare, had occasionally knocked out a radio, or caused minor structural damage, but had never been known to take down a plane. Since the investigators of this tragedy had little else to go on they began to investigate the fire of St. Elmo a little more closely.

After ruling out the list of usual suspects, the investigators reexamined the results of metallurgical tests and determined that the fuel tanks had undergone extreme pressure and had exploded causing the wing of the aircraft to rip away. They believed that St. Elmo's fire had entered a pressure vent and ignited gas fumes which led to the explosion of the fuel tanks. Lockheed Aircraft, who had built the aircraft performed a reenactment of the event at their proving facility in California. They built a replica of the plane's wing and attempted to duplicate the event of that fateful June day. Their experiment succeeded, and Lockheed then ruled that the cause of the crash was due to an errant case of St. Elmo's fire.

Examining Lockheed's conclusions I find it hard to believe that the fire of St. Elmo was the definitive culprit. In order to prove this beyond a shadow of a doubt, the atmospheric conditions related to the experiment must be exactly the same as they were the day of the crash. Also, like ball lightning and the will-o'-the wisp, the fire of St. Elmo, although better understood now than in centuries past, still possess' a lingering aura of mystery around it. How can it ignite a fuel tank on an airplane, but travel harmlessly up a man's sleeve, or appear on the tip of an umbrella without burning the fabric below it? These questions need to be answered before the fire of St. Elmo can be convicted for causing the TWA disaster.

Probably the most famous example for St. Elmo's fire being blamed for causing an aircraft disaster did not involve an airplane at all. On May 6, 1937 the German Zeppelin Hindenburg was nearing the completion of its first North American run of the 1937 season. The Hindenburg was a massive airship over 800 feet long built from the finest material then available. It was the pride of Germany, which, of course, at that time was run by the Hitler led Nazi machine. The huge dirigible had had an uneventful run. However, as the great airship was about to connect with the mooring mast at the Lakehurst Naval Air Station in New Jersey, it suddenly, inexplicably, burst into flames. A 32 year old radio announcer named Herbert Morrison immortalized the disaster in a choke filled broadcast which was later heard all around the world. In the end, 36 people perished in the inferno while miraculously 61 people managed to escape from the carnage.

The Hindenburg disaster was first thought to have been the work of saboteurs. However, after carefully examining the wreckage, and the blueprints of the airship, it was determined that the great ship was doomed from the start. The disaster was blamed on hydrogen. In its true form hydrogen is not flammable, but when mixed with oxygen can be frightfully so. The Hindenburg was carrying at least seven million cubic feet of hydrogen which was being used primarily as a lifting gas. The ship was also carrying plenty of diesel fuel. The combination of hydrogen and diesel made for a deadly potion that inevitably burned the ship. No one knows exactly what caused the fire to start, but it is known from photographs and eyewitness accounts to have started in the rear part of the aircraft near the upper fin. Some people later reported seeing a bluish glow radiating from the upper surface of the Hindenburg shortly before the fire broke out. Of course, this bluish glow is the kind of light normally associated with St. Elmo's fire.

One of the German investigators who attempted to find a cause for the Hindenburg disaster was a physicist named Max Dieckmann. He believed that because of the recent thunderstorm in the area around Lakehurst there would have been a great difference in electrical potential between the clouds and the ground beneath the Hindenburg. As long as the ship was airborne it would not have produced a problem. However, once the first mooring line came in contact with the wet field, it grounded the ship. After more mooring lines were added to the mesh, the ship soon acquired the same charge as the earth below it. The electrical current which discharged into the atmosphere then ignited the hydrogen. According to Dieckmann, this was a recipe for disaster.

Is it possible that the bluish glow seen just before the airship caught fire was the fire of St. Elmo? It might have been, but it is just as likely that the fire was caused by something else. According to Zeppelin managing director Hugo Eckener when the Hindenburg was making its final approach it made a sharp turn which could have caused a bracing wire to snap which would have ripped an opening in one of the hydrogen gas cells. Whether or not the fire of St. Elmo caused the spark that ignited this hydrogen is debatable. It could just as well have been caused by some careless crew member lighting a cigarette. Whatever the cause, the fire of St. Elmo will always go down in history as being either a savior as the name implies, or a culprit as some recent history has thought to deem it. Scientifically, it is merely one of our planets many wonders.

Chapter 14

THE FIREBALL OF THE MAGI

In this chapter I will attempt to show that one of the most talked about figures in history was born under the majestic glow of a fireball. Of course, the figure that I refer to is none other than Jesus Christ of Nazareth who hardly needs an introduction here. For two-thousand years the story of Jesus has been told countless times in many languages. In the western world he has no equal. The religion based on his teachings reaches out to over a billion people worldwide. The birth of Jesus has been the subject of much debate since scholars first thought to study his life. It is said that he was born under a star, and the examination of this is what this chapter will be all about. However, first I must mention that Jesus is by no means the only religious figure to have been born with the aid of a star. One only has to look east, and examine the origin of Buddha to see that the star motif is visible in that part of the world as well. Let us look at the following verse written by Sir Edwin Arnold:

> That night the wife of King Suddhodana,
> Maya the queen, asleep beside her lord,
> Dreamed a strange dream; dreamed that a star from heaven-
> Splendid, six rayed, in colour rosy pearl,
> Whereof the token was an elephant
> Six tusked, and white as milk of Kamadhuk-
> Shot through the void; and, shining into her,
> Entered her womb upon the right, Awaked,

Bliss beyond mortal mother's filled her breast,
And over half the earth a lovely light
Forewent the morn. The strong hills shook; the waves
Sank lulled; all flowers that blow by day came forth
As twere high noon; down to the farthest hells
Passed the queens joy, as when warm sunshine thrills
Wood-glooms to gold, and into all the deeps
A tender whisper pierced. "oh ye," it said,
"The dead that are to live, the live who die,
Uprise, and hear, and hope! Buddha is come!

This verse is only part of the great epic poem *The Light of Asia* which was penned by Arnold in the mid 19th century. The poem was a bestseller at the time for Arnold who got his inspiration from the *Buddha Charita* written by Asvaghoska a 1st century Buddhist monk. The *Light of Asia* details the history of Prince Siddartha, who later became the Lord Buddha. Arnold formed his poem into eight books. Book one deals with Siddartha's birth until he reaches the age of eighteen when he begins his long journey and transformation into the Buddha. From the verse above it is clearly seen that the birth of Siddartha is entwined with that of a fireball. We must remember however, that although the poem itself contains the artistry and poetic license of the author, the poem itself is an accurate adaptation from earlier sources, namely the *Buddha Charita*. The facts, or the bulk of the story, however, is the same. Maya, the mother of Siddartha has a dream of a "star" or "light" from heaven. This star shoots through a void and enters her womb. Half the Earth then experiences a lovely light and the strong hills shook. The Buddha is born.

This "light," present at the birth of Siddartha was almost certainly a meteor. It was probably something along the line of the Sikhote-Alin meteor of 1947. Something this big coinciding with the birth of a prince would have been talked about and the parallels would have been too close not to notice. It is said that Buddha was born sometime around 560 B.C. but this date is not engraved in stone so that there is really no way of correlating a specific event from the Chinese or Indian records with the date of his birth.

So, seeing proof that the Buddha was born under a "star" we now turn to the main focus of this chapter, the birth of Jesus Christ. The story is well known, and its various details have been studied and examined by scholars for centuries. However, like most of the Bible, part of it is wrapped in riddle, and needs to be looked at closer. Like the older story of Siddartha, here also is a major religious figure whose birth is entwined with that of a fireball.

Let's us see how this comes about by citing scripture. Matthew 2:1:2:

> When Jesus was born in Bethlehem of Judea, in the
> Days of King Herod, behold, magi from the east
> Arrived in Jerusalem, saying, "where is the newborn
> King of the Jews? We saw his star at its rising an
> Have come to do him homage.

Before analyzing this verse let us skip ahead to Matthew 2:7:11:

> Then Herod called the magi secretly and ascertained from them
> the time of the star's appearance. He sent them to Bethlehem
> and said, "go and search diligently for the child. When you have
> found him, bring me word, that I too may go and do him homage."
> after their audience with the king they set out. And behold, the
> star that they had seen at its rising preceded them, until it came
> and stopped over the place where the child was. They were
> overjoyed at seeing the star, and on entering the house they saw
> the child with Mary his mother. They prostrated themselves and
> did him homage. Then they opened their treasures and offered
> him gifts of gold, frankincense and myrrh.

A lot has been said about the "star" associated with the birth of Jesus. One common assumption which has found itself caught into the web of Christian tradition is that this star that the Magi were following was the North Star, and that it was the brightest star in the sky. Nothing can be further from the truth. The brightest star in the sky is Sirius with a—1.5

magnitude. In fact, this star is so bright that to the armchair stargazer it is often confused as a planet. The North Star, known as Polaris has only about a 2.0 magnitude. In other words, sometimes, even on clear nights you still have to look for it, and know exactly where to look. Still, even with this bit of information it would not be totally out of the realm of possibility for the North Star to be the magi's star if it were not for one piece of damning evidence that proves it was not. The Magi met with Herod in Jerusalem, and after taking their leave set out toward Bethlehem. Bethlehem is south, not north of Jerusalem. Since this is the case, if the Magi were following the North Star they would have been heading in the wrong direction. Polaris was behind them as they journeyed to Bethlehem.

We have now shown that Polaris was not the star that the Magi followed. So if it was not, than what star could they have been following? Take your pick. It is possible that they could have been following any one of a number of bright stars visible in the southern sky at that time of year. This would include Sirius. However, examining scripture a little closer tells us that it probably was not one of the so called misnomers called "fixed" stars.

Matthew 2:2 says that the Magi have an audience with King Herod and tell him "we saw his star at its rising." The key word here is "HIS" which signifies something personal. "HIS STAR" is different than saying "THE STAR." The "HIS" therefore means that the Magi were obviously referring to a star that was newly arrived. After all, it would not make any sense for the Magi to look out the window one night and peer up at the sky and point to Sirius or Rigel and say "look there is "HIS" star." These stars are there every night, visible in the northern hemisphere for a good part of the year. To the avid stargazer there is nothing special or extraordinary about them. They would not assign any specific importance to these heavenly bodies. So if it was not one of the brighter stars that the Magi were referring to than what was it? At this point in our discussion there are four possibilities. The Star of the Magi could have been a supernova, a comet, a conjunction of the planets, or . . . dare I say it . . . a blazing fireball!

Before examining these four possibilities we must ask ourselves an important question. Who were the Magi? This question has been the subject of much debate and conjecture on the part of scholars for many years. It is often assumed that the Magi numbered three. Indeed, the Scribner-Bantam English dictionary defines "Magi" as "The three wise men who came from the East to see the child Jesus." Although the number "3" has generally been accepted as the number of Magi present at the birth of Jesus there is no evidence that this was their actual number. It is only an assumption that there were three Magi present at the birth of Jesus in Bethlehem. This, of course, is based on the fact that there were three gifts; gold, frankincense, and myrrh, but this is hardly evidence to lead to the conclusion that three was the actual number. We can only conclude that there were more than one Magi that visited Jesus due to the fact that Matthew refers to the Magi in the plural a number of times using the plural pronouns "we", "they" and "them." So we can safely say that there were at least two of them, but who were they? According to the Saint Joseph edition of the New American Bible the Magi was a term that had been originally used to denote a person of the Persian priestly caste. They were often said to possess a vast amount of knowledge. In those days, before trained scientists were used to establish a methodology, and inquire into the nature of how things functioned and worked, the interpretation of the various branches of science was generally left to the priestly caste. Most of their interpretive knowledge came about by observing the movement and various phases of the sun, moon, planets, and stars. They were astrologers, and it was common for them to construe a celestial event with things like the birth of a ruler, or the movement and fate of an army. Kings would often use them as sagacious fortune tellers. One only has to look at the semi-mythical sage known as Merlin who is said by Geoffrey of Monmouth to have been instrumental in the birth of the Saxon King Arthur Pendragon. Merlin is often portrayed wearing a robe of satin covered with stars which is the symbol of his craft. Indeed, throughout history astrologers have influenced great rulers, and can said to be responsible for certain events that have shaped the course of history. Over the past few centuries a number of these "wise men" have appeared on the world's stage. One of them was a supposed monk named Rasputin, who, though a commoner, was able to have a heavy influence on the imperial court of the Russian Czar Nicholas II. It is often said that Rasputin was a key ingredient in bringing about the Russian Revolution which brought an end to imperial Russia. There is also the celebrated case of the enigmatic Comte de Saint Germaine who lived in 18th century France, and was said to have been one of the instigators of the French Revolution. Saint Germaine was known to have dabbled in astrology, and although he was supposed to have died in 1784, sightings of the Count were reported well into the 19th century leading to speculation that he was immortal. The cases of Rasputin and St. Germaine are stories worthy of a book themselves, and as fascinating as they are I mention them here only to show the parallel between them and the Magi of biblical tradition. Placing them in context with today's world they would in all likelihood be regarded in this day and age as nothing more than charlatans. Still, they would have been familiar with the stars and it is quite likely that they would have demanded a great deal of respect because of their talent. If some unknown quantity was seen in the sky they almost certainly would have been called to duty to interpret it's incongruous nature and meaning. There is a good possibility that this was the manner in which the Magi first received word of the fireball seen over Bethlehem sometime around 7 B.C.

So, although there is no way of knowing exactly where it was that the Magi were from it is possible that they came from Persia. The purpose of their journey according to Matthew was to pay homage to the baby Jesus. However, upon arriving in Jerusalem it seems obvious that they do not know where this new born king is, for they ask Herod, "Where is the newborn king of the Jews?" Herod, of course, is naturally dumbfounded by this question and has to confer with his scribes and priests who tell him that the Messiah will be born in Bethlehem as the prophet had written. Herod then relates this information to the Magi who set out for the place. It is unclear, however, if the Magi asked Herod this question to see if he was familiar with the prophesy or whether they were only familiar with part of it and did not know themselves. Then again, if they had information regarding the fireball, and also knew that it landed somewhere in Bethlehem, they were merely asking Herod for directions to a town that they themselves had probably never visited.

Since the Magi were astrologers it is quite possible that they had come west armed with scant information on the fireballs terrestrial landing place. It can be assumed that they had some sort of reason for believing that it had landed somewhere in Judah, but were not privy to its actual place of fall. I have mentioned throughout the course of this book the notion of fireballs being related to the birth or death of great rulers. Caesar's soul supposedly left the earthly plane riding a comet, and Buddha was born at the time when a spectacular "rosy pearl" fireball shot through the heavens. Since fireballs are quite common occurrences, every one seen and reported to the Magi could not possibly have been the sign heralding the future reign of a king. Therefore, this fireball over Bethlehem must have been a spectacular and magnificent sight. The Magi must have received reports which gave them cause to believe that this was a "new star" which was worth checking out. The Magi, possibly with future trade relations in mind set out on their quest which was probably secular and politically motivated. Why else would these Persian priests travel to another land to pay homage to a future Jewish king? I cannot think of any other reason except for the slim possibility that the Jewish scribes like Matthew started to refer to their priests or astrologers as "Magi" in the Persian sense of the name. So it is quite possible that the Magi were not Persian at all, but Jewish and hailed from the eastern region of Israel perhaps somewhere east of the Jordan river.

Now, another important question arises as to how the Magi and scribes of Herod knew that the Messiah was to be born in Bethlehem. Indubitably, they were well read, and had studied the (even then) ancient scrolls. In the Old Testament the book of Micah makes it abundantly clear. Micah 5:12 reads:

> But you, Bethlehem-Ephrathah
> Too small to be among the clans of Judah,
> From you shall come forth for me
> One who is to be ruler in Israel;
> Whose origin is from old,
> From ancient times.

Therefore the Lord will give them up, until the time
When she who is to give birth has borne,
And the rest of his brethren shall return to the
Children of Israel.

Basically, the essence of this prophesy states that a child descended from ancient lineage was to be born in Bethlehem, and that this child will one day be ruler of all Israel. The ancient lineage meaning one who shall have descended from David.

It is said that Micah lived roughly around the time that Isaiah was writing which if this be the case would probably have been around 725 B.C. This would tell us that a period of over 700 years had elapsed from the time of this prophesy until the time of the birth of Jesus at Bethlehem. This, in itself, is further proof that this fireball must have been an awesome sight to behold. Surely it must have been in order for it to have the prestige of fulfilling a prophesy written seven centuries earlier.

We have now been able to ascertain who the Magi might have been and what their purpose was in traveling to Bethlehem. I earlier stated four possibilities that the Magi's star might have been; a supernova, comet, conjunction of the planets, or a fireball. I am of the opinion that it was a fireball, but before I bring forth the evidence which supports my position I must first debunk the claims of the other three.

The possibility of a comet being the Magi's star is not new. In fact, the idea has been floating around since at least the 14th century when the Renaissance painter Giotto di Bondone created one of the most famous images of that era. The painting is called *Adoration of the Magi*. The painting, on fresco, shows the baby Jesus surrounded by the usual suspects that are associated with his birth. Mary, Joseph and the Magi hover around the child who is protected from the elements only by a humble looking wooden structure that is not even protected by walls. Indeed, it appears much like a modern day carport made of wood. Above this place of refuge, visible in the night sky, is a comet. Indeed, according to Encyclopedia Britannica, Giotto observed Halley's Comet on its passage in 1301. This event most probably influenced Giotto's thinking when he painted the *Adoration of the Magi* only a few years later.

Giotto may have believed that the Star of the Magi was a comet, but there are a few good reasons why I believe that it was not. First, a comet, visible in the sky is an intermittent object and might cause some excitement when first viewed, but would have had no personal affiliation with the town of Bethlehem. In all likelihood the comet would not only have been visible there, but in much of the northern hemisphere as well. Secondly, as I have previously mentioned, a 700 year block of time had elapsed from the prophesy of Micah to the birth of Jesus. There would have been countless comets visible in the northern hemisphere during that time period. Indeed, Comet Halley itself would have made at least 10 journey's around the Sun during that time. Not to mention all of the non-periodic comets that would have passed by the Earth during this time. With this being the case, would not a previous comet have had the Magi or their forbears trekking to Bethlehem in search of the future king? Surely the great comet of 130 B.C. which Kirkwood called "The comet of Mithridates" should have had someone heading in the direction of Bethlehem? This comet was said to have been as large as the Sun. Donald K. Yeomans mentions another bright comet in his book *Comets, A Chronological History of Observations, Science, Myth, and Folklore*. This comet was seen in 147 B.C. and recorded by the Chinese as well as the 1st century Spanish/Roman scholar Seneca who wrote that the comet was as large as the sun, reddish in color like fire, and bright enough to dissipate darkness. These two comets seen in the century before the Magi's star are just two examples of the many comets recorded by observant scribes from many countries in the centuries before Jesus' birth.

Based on these two assumptions alone it would seem that a comet could not have been the star that the Magi were following. However, a third reason just about puts a nail into the comet theory's coffin. This has to do with the average persons superstitious notions of comets in those days. Comets were looked on with trepidation in the belief that after one appeared, some sort of pestilence or war would erupt. Indeed, the fear of comets has survived even to recent times. Nearly 1500 years after the birth of Jesus, in 1456, a comet appeared in the heavens. This comet just happened to coincide with the fall of Constantinople into the hands of Muslim Turks. Sitting in Rome, the elderly Pope Callistus III saw this as a bad sign, and one that might have had a direct outcome of the city's surrender. The comet was described as "red and hairy" and the Pope must surely have looked at this with horror. He immediately called for several days of prayer and fasting, believing that this would cure all the evils that had befallen on the city and it would once again be returned into Christian hands. Callistus was disappointed when this failed to produce the desired results. Desperate, and lacking the resources of launching a full fledged crusade to take back the city, he decided that he had only one option left. He excommunicated the bloody red comet by issuing a Papal Bull! Perhaps somebody today can petition Pope Benedict to lift the burden off this poor icy celestial body known today as Halley's Comet.

We have now determined that a comet was not the Star of the Magi. Let us now look at the possibility of a supernova. This is a rare event indeed. At least it is in our part of the universe that is visible to the naked eye of an observer on Earth. Throughout the course of recorded human history there have been only a few instances of one of these catastrophic events on record. In actuality, across the universe, they are a lot more common than one might think, but it just so happens that they do not occur that often in our part of the Milky Way. What is a supernova? In lay terms it is a star that becomes so heavy that it eventually collapses under its own weight. The nuclear reactions within the star after its collapse cause the temperature to soar to over 100 billion degrees at its core. Heavier elements like iron can be produced at these high temperatures. Finally, the star explodes and violently spews out its material into outer space at about 1000 miles per second.

Probably the best known supernova on record is the one that took place in the year 1054. The event was dutifully transcribed for posterity by Chinese scholars. Of course, at this time in history they could not possibly have known what it was that they were witnessing. The remnants of this supernova can still be seen today through a small telescope. It is now known as the Crab Nebula named for it's resemblance to a crab.

It appears as a hazy cloud of dust. Eventually, some of this dust may find itself drawn into a stellar nursery where a star is being formed. This ancient material is thought to contain the building blocks of life, and it is generally believed that our solar system formed from this primordial material sometime around 5 billion years ago.

One can only imagine what the Chinese astronomers of that day thought when suddenly a strange beacon of light appeared in the heavens where none had been before. It would have been the talk of the town. Is it possible then, that the Magi could have been following a supernova that appeared sometime around 7 B.C? Possible . . . but doubtful, and there are a few reasons why I believe this. First, like a comet, a supernova would not have had a personal relationship with the town of Bethlehem. It too would have been visible all over the northern hemisphere. Secondly, the records of this period remain silent. If an event of this magnitude had occurred surely Seneca, Pliny the Elder or some other conscientious chronicler would have made mention of it in the generations following the birth of Jesus. However, there is no record, not even in China who it may be said, were, and are, meticulous keepers of records. Because a supernova is such a spectacular event, and it is so rare, I find it hard to believe if one occurred around 7 B.C. there would not at least be a passing mention of it somewhere. It is because of this that I do not believe the Magi's star was a supernova.

The next possibility is that the Magi's star was a conjunction of the planets. A conjunction occurs when two planets line up on the same ecliptic plane. When this happens, the planets reflected light amalgamates into one strong beacon which the viewer on Earth perceives because he is in line with it. To the observer on Earth the reflected light would be so bright that it might appear that a new star had suddenly appeared in the heavens. Like the comet theory, this idea that a conjunction of planets was the Magi's star has been floating around for centuries. It first gained credence in the early 17th century when Johanne Kepler calculated that a conjunction of the planets Saturn and Jupiter occurred in the year 7 B.C. In fact, not only did it happen once that year, but twice more. This is called a triple conjunction. These are rare events and it is easy to assume that the Magi would have read some sort of divine intervention here. However, and I will repeat this for the last time, a conjunction of the planets would not have had a personal relationship with Bethlehem. It also does not explain the Magi seeing the star in the east, and then losing sight of it before seeing it again over Bethlehem. Now . . . on first appearance it might seem that this would be a possibility, for according to astronomical calculations the conjunctions would have occurred during the months of May, October, and December of the year 7 B.C. However, it must be remembered that Herod asked the Magi when it was that the star first appeared. We know from Herod's reaction and subsequent brutality inflicted on the young boys of Bethlehem, that up to two years had elapsed from the time of Jesus' birth to the Magi's visit. For it is written, Matthew 2:16:

> "When Herod realized that he had been deceived by the Magi, he became furious. He ordered the massacre of all the boys in Bethlehem and its vicinity two years old and under, in accordance with the TIME he had ascertained from the Magi."

From this passage we can safely deduce that Jesus, at the time of the Magi's visit to Bethlehem was probably around two years old. This would make the revised date of the Magi's visit to Bethlehem to be around 5 B.C. Certainly no later than 4 B.C. because that was the year of Herod's death. Why it took the Magi two years to reach Bethlehem is a mystery. Perhaps they were busy performing calculations on other astrological conundrums. More than likely, their visit to Bethlehem was nothing more than a formality.

We can now plainly see that a conjunction of the planets; Jupiter and Saturn could not have been the Star of the Magi. The time frame just does not fit. The conjunction could have accounted for the first sighting of the star, but it could not have explained the second. There would also had to have been a conjunction in the year 5 B.C. for this theory to hold up. There is no record of a conjunction of Jupiter and Saturn occurring in that year. In order for this theory to join the ranks of plausibility one would have to admit an error attributed to Matthew regarding the linear placing of events.

Our attention can now focus squarely on the most plausible theory of the nature of the Star of the Magi. We will start with their journey from the East. Let us assume for the sake of simplicity that one crystal clear night one of these "wise men" was standing in the portico of his home gazing contemplatively up at the heavens. This, of course, was way before the advent of the telescope so the Magi's view of the sky was wholly peripheral, and his gaze probably wandered this way and that. He might have even been admirably studying the conjunction of Jupiter and Saturn when behold! Something in the eastern sky drew his attention in that direction. The whole sky seemed to brighten and when he turned his head to get a better look he was treated to an amazing sight. There in the sky was a dazzling spectacle. A resplendent red manifestation that quivered slowly but methodically across the Lord's firmament. This Magi undoubtedly had seen fireballs before, but he had never been witness to one like this. It lit up the night sky many times brighter than the full moon. Causing eerie shadows to dance across the arid Persian landscape. The "wise man" was stunned, his mouth agape and his animated countenance would have seemed ludicrous to anyone who happened to be looking at him. But that was not the case. People stood in their doorways, or peered out of windows, while some filed out into the street for a better look. All of them gazing upward, transfixed by this wonder from the heavens that was as large as the sun and just as bright. Some people panicked, perhaps believing that the end of the world was at hand.

As we continue our hypothetical tale, I will remind the reader that it is not certain where the Magi hailed from. It is only known that they came from the East. Although I believe that they might not have been Persian at all, let us say for the sake of argument that they came from the cradle of civilization along the banks of the Tigris and Euphrates rivers in a little village outside of modern day Baghdad. It is quite possible that our Magi along with countless other people could have heard the rumbling as the fireball broke up. After all, the Siberian explosion of 1908 was said by some to have been heard at distances up to twice that distance. I am not suggesting that this fireball was as deadly as the Tunguska fireball. It was probably more in line with Sikhote-Alin which was also heard from a considerable distance but was nowhere near as destructive as Tunguska.

We will return to our Magi in a moment, but let us first look at another piece of crucial evidence that is necessary to relate which supports the fireball theory. A short time after the fireball was seen streaking across the sky over Persia, a shepherd was tending to his flock of sheep in the fields outside of the gates of Bethlehem. The night must have seemed like any other to this humble man. However, like the Magi standing in the portico of his house 500 miles to the east . . . that would soon change. It was late in the evening as the elderly shepherd ambled his way down to a bubbling brook to fetch a refreshing drink of water to cool his parched throat. He used his staff to assist him as he squatted down and dipped his ladle into the chilled water. Then something happened. He perceived a strange light out of the corner of his eye and he instinctively turned his head in that direction. The blinding glare forced him to squint. He shielded his eyes with his forearm and accidentally dropped the ladle filled with water before he could even get a sip. The whole sky gleamed with a reddish-orange glow. Suddenly there was a loud popping sound followed by a thunderous roar as the fireball exploded into thousands of fragments. The good shepherd watched in awe as this blazing thunderstone of the gods ended its four-billion year journey through the solar system, by shooting out fiery sparks like a 4th of July firecracker before expiring in a resonating final boom.

For a few minutes after this hellish apparition there was a dead silence. Not a soul moved from where they were standing when the fireball appeared. They must have wondered if the show was really over. What was going to happen next? Would the gods send forth a tempest that would annihilate their entire civilization? What grievous sins had they committed to warrant this?

After a few minutes the trepidation of the shepherds was mitigated somewhat, and slowly their courage returned to them. Especially when they began to gather in numbers in a pasture outside of the town. Under the moonlight each one of them related their experience, and gained a sort of comfort and guidance from the more learned of their lot. It wasn't long before word got around that a child had been born the very moment that this prodigy had been seen. At first there was a quiet murmur among the shepherds and other tradesmen and travelers who had gathered en-masse and had now made their way to the gates of the town. It wasn't long before a priest joined the ensemble. He climbed up on a large rock and began to address the crowd which by now had become quite boisterous. With a few loud but soothing words the priest silenced the multitude who had become quite anxious for someone of authority to interpret the meaning of the fireball.

"Good people!" bellowed the priest "Yahweh has spoken!"

The crowd of shepherds, merchants, and travelers began to mumble and whisper among themselves. The priest, who was new to the town himself saw the fear and distress in their countenances which were eerily illuminated by the scores of torches around them.

"Does this portend some disaster? . . . a famine? or a pestilence perhaps?" queried one nervous shepherd.

The priest raised his arms in an attempt to calm the frightened throng.

"Do not fear! . . . no calamity will come of this! . . . this is a time to rejoice, for on this night a savior was born who will one day rule over all the children of Israel! How do I know this? . . . because it is written by the prophet!"

We will turn back to the adventures of the Magi and the shepherds, but first we need to clear up one common misconception. That, being the notion that the Magi's star was only recorded in the Book of Matthew. This is not the case. The truth of the matter is that it was also subtly mentioned in the book of Luke. Let us look closely at this. Luke 2:18-14 reads:

> Now there were shepherds in that region living in the fields and keeping the night watch over their flock. The angel of the Lord appeared to them and the glory of the Lord shone around them, and they were struck with great fear. The angel said to them, "do not be afraid; for behold, I proclaim to you good news of great joy that will be for all the people. For today in the city of David a savior has been born for you who is Messiah and Lord. And this will be a sign for you: you will find an infant wrapped

in swaddling clothes and lying in a manger. And suddenly there
was a multitude of the heavenly host with the angel, praising
God and saying: Glory to God in the highest and on Earth peace
to those on whom his favor rests.

These biblical passages are well known, but not well understood. Luke
is being elusive, and speaking in riddle. The reason for this is unclear.
Possibly to hide the truth from would-be enemies that would suppress
any dissident behavior. A whole book could be written on why a lot of the
Bible is written in this manner. Even Jesus talks to us in parables. With
this being the case, a careful analysis of these passages tells an amazing
story even if at first glance it appears to be a forthright assessment of what
went down. By all outward appearances it looks as if the shepherds are
visited by an angel from heaven. At least this is the way that it is portrayed
for it says Luke 2:9:

The angel of the Lord appeared to them and the glory of the
Lord shone around them, and they were struck with great fear."
Is it possible that the angel of the Lord was only a traveling
priest or other person of high standing that came from inside
the town to reassure these humble people that they had nothing
to fear from this message from God?" Is it also a possibility
that the "glory of the Lord shown around them" was nothing
more than a meteor hurling through the Earth's atmosphere?
Not knowing what this fireball was, and having no concept that
rocks could fall from the sky, the interpretation of the author of
the book of Luke is in good faith considering that he lived many
years after the events related in these passages. Luke is said to
have been a Syrian who wrote his gospel around 50 years after
the death of Jesus. He obviously received his information from
another source, which, unfortunately for us, is now lost. There
is a plausible explanation for the shepherds believing that the
informant of this news was sent from God. The traveling priest
or dignitary who would have been an unfamiliar face to the
shepherds could account for this. Bethlehem was only a short

distance from Jerusalem and would have been a resting place
for weary travelers on the road to and from Egypt.

After reassuring the shepherds that all was well, and that the prophesy
of Micah had been fulfilled, we come to a passage in Luke that is puzzling
and seemingly does not quite fit with the rest of the text. It seems to have
been inserted there, and I can think of no other reason for its placement
except perhaps that the author, or some translator over the course of many
centuries did not understand it. The passage that I am referring to is Luke
2:13-14 which says:

> "and suddenly there was a multitude of the heavenly host with
> the angel, praising God and saying "glory to God in the highest
> and on earth peace to those on whom his favor rests."

Reading this and forming a mental image of it, you would think that the
"angel" was soon joined by other angels and together they said a prayer to
Yahweh. The "multitude of the heavenly host" is interpreted by Christian
scholars to mean that other angels soon joined the initial messenger. At
least this is how it was translated. I ask a simple question . . . Could
it mean something else? Could the "multitude" refer to fragments of a
stone which is then turned into "the heavenly host?" Now this is a strange
choice of words, for it is common knowledge that the host is another name
for the Eucharist which, of course, is symbolic of the body and blood of
Christ. However, this word also has another meaning. The word "host"
translated into Hebrew is "Sabaoth" which implies that it is an "army of
something." Let us look at the book of Psalms 33:6

> "By the Lord's words the heavens were made;
> By the breath of his mouth all their host"
> The "host" in this passage refers to an "army of stars."

Let us look at the book of Jeremiah 33:22

> "Like the host of heaven which cannot be numbered, and the
> sands of the sea which cannot be counted, I will multiply the

descendants of my servant David and the Levites who minister
to me."

These words were told by God to the prophet Jeremiah when he was imprisoned by King Zedekiah of Judah for prophesizing. It is clear here that "host" is meant to denote a number of stars. It can now be plainly seen that the "host" mentioned in Luke could also refer to an "army of stars." Or, better yet an army of FALLING STARS. There are other places in the Bible where this term is used to refer to stars in the sky, but with this in mind let us turn back to the shepherds who have now made their way into the town proper to bear witness to the object of veneration that they have been told is "wrapped in swaddling clothes, and lying in a manger." Luke 2:15-20:

> When the angles went away from them to heaven, the shepherds said to one another, "let us go, then, to Bethlehem to see this thing that has taken place, which the Lord has made known to us. So they went in haste and found Mary and Joseph, and the infant lying in the manger. When they saw this, they made known the message that had been told them about this child. All who heard it were amazed by what had been told them by the shepherds. And Mary kept all these things, reflecting on them in her heart. Then the shepherds returned, glorifying and praising God for all they had heard and seen, just as it had been told to them.

Upon initial inspection of these passages it would seem that the shepherds went to the manger to pay homage to the future king. In all probability this is what they had gone there to do. However, there is something puzzling about the reaction that Mary and whoever else was with her had when this information was revealed to them. Why were they amazed by the shepherds revelation? Luke never divulges any particulars about the message, but it can be inferred that at least part of it was the prophesy told to them by the angel or traveling priest that they had spoken with outside the walls of the town. For it says "They made known the message that had been told them about this child." We know that the angel/

priest informed the shepherds of the prophesy that had been fulfilled. But why would this amaze Mary, and why was she only reflecting on it after the shepherds had arrived and told her about it? She would in no way have been surprised to hear this information, for it had already been told to her. This does not make any sense, and to me contradicts passages in other gospels, for not only had the angel appeared to Mary as we all know, but also appeared to her husband Joseph in a dream. They, at least, were aware of the child's divinity. Certainly Mary would have been, as she had had no relations with any man. So . . . once again, I ask the question . . . Why were they amazed? The gospel of Luke remains silent on the matter but indubitably it was something of importance or Luke would not have mentioned it. Perhaps in Luke's day when Christianity was still an outlawed religion this mystery was well known and he might have thought that it was unnecessary to disseminate any of the details. It appears that at this late date one can only speculate. Could the shepherds have told her of a spectacular fireball that had lit up the sky the moment of Jesus' birth? This information might have had the effect on Mary that Luke tells us. One thing is certain . . . whatever it was that the shepherds had divulged to her she had not yet heard.

We are now at the end of the shepherds involvement regarding the birth of Jesus. We must now return to the Magi who if you recall had just witnessed the fireball streaking across the Persian sky many miles to the east of Bethlehem. After the initial excitement had worn down, the Magi must have immediately set to work trying to figure out the significance of this portent. They must have consulted each other, searching scrolls and ancient tablets. Runners were sent out to gather information, and eventually word got back to them that the fireball had come to rest in Bethlehem. Someone, perhaps a diplomat familiar with Jewish custom and tradition, or indeed a member of the tribe of Israel himself filled in the final pieces of the puzzle. It took a while, but a delegation was finally sent out and reached Jerusalem to meet with Herod as we have already related. We have already ascertained that up to two years had passed from the time of the stars first appearance until the day that the Magi entered Bethlehem. The question that we must ask now is, why did Joseph and Mary decide to stay in Bethlehem? One would think that they would have returned to Nazareth. According to the book of Luke, the only reason that they went to Bethlehem in the first place was because of a decree issued by the Roman Emperor, Caesar Augustus that required each man to return to the town of his birth to take part in a census. Now I see a few possibilities as to why they might have decided to stay. Perhaps the government agents taking the census took a while to reach Bethlehem. This is a possibility, but a more likely scenario was that Joseph had found work here and had decided to stay for a while. We will probably never know the reason, but we do know that when the Magi arrived, Jesus and his parents were still in town and the star had reappeared.

Illustration 16

Matthew 2:9-11 reads:

> After their audience with the king they set out. And behold,
> the star that they had seen at its rising preceded them, until it
> came and stopped over the place where the child was. They were
> overjoyed at seeing the star, and on entering the house they saw
> the child with Mary his mother . . .

So we see from the passage above that the star had clearly disappeared, and the Magi did not see it again until they reached Bethlehem. It is my opinion that what the Magi actually saw when they reached Bethlehem was a meteorite. In other words it was the star in its terrestrial form much as the Son of God had transformed from spirit to flesh and blood. The meteorite, therefore, I believe, was set up over the residence of the future King of the Jews as a symbolic gesture that suggested the symbiotic relationship between heaven and Earth. Possibly, the rock on display was the largest fragment of the meteorite. Indeed it is also possible that this was what surprised Mary so much after her conversation with the shepherds. Perhaps she had harbored some doubts about the angel's revelation and this was the solid evidence that was needed to "reflect on them in her heart."

One can now picture the delegation of Magi arriving in the town. Word had already reached Bethlehem that they were coming. The reason for their visit might not have been made known to Joseph and Mary until word arrived from Jerusalem that they were on their way to visit after their meeting with Herod. We can picture one of the shepherds guiding the delegation into town through the north gate. For the sake of simplicity let us say that the Magi numbered three. The entourage would have created a convivial, carnival-like atmosphere. The Magi wearing robes of crimson and scarlet sat stoically in a horse drawn coach adorned with velvet curtains which were fashioned in the Persian style. The coach ambled its way down the dusty street leaving a cloud behind it. A young boy standing in a doorway somewhere along the route noticed the curtain part slightly. A hoary head exhibiting sagacity peered out at the throng of bystanders watching the carriage pass. As the carriage made its way down the main street a crowd of curiosity seekers followed in its wake. The carriage turned down a narrow lane and stopped in front of an inn. Slowly the three "wise men" climbed down from the coach and studied the place where they had stopped. A priest emerged from a doorway of the inn and approached them with an outstretched hand beckoning them to follow him into the courtyard. The delegation passed through a wooden gate and found themselves standing in front of a raised platform in which a spiral staircase constructed from the hands of Joseph himself twisted its way to the top. The priest led the way up the stairs and motioned for the Magi to follow. When they reached the top they were awestruck by the sight in front of them. Matthew 2:9: "And behold, the star that they had seen at its rising preceded them, until it came and stopped over the place where the child was."

Here it was, in all of its glory, the symbol of heaven that had brought forth the coming of the king foretold by the prophet Micah centuries before. In a large reed basket carefully placed on the raised platform was a large black stone . . . the star of the Magi in its terrestrial form . . . an iron meteorite. The Magi would have carefully examined this relic and then asked to see the child and his mother. Climbing down from the platform they made their way into the inn where Mary received them warmly. After paying homage to the future king and leaving the three gifts of gold, Frankincense, and Myrrh they took their leave, and with this they disappeared from history. It is unknown how long the Magi stayed in Bethlehem. It is probable that they at least spent the night. Where they went after their visit is unknown. They could have gone back to Persia, or perhaps traveled on to Egypt on business. It is unlikely that they had made the trip to Bethlehem for the sole purpose of paying reverence to a future king. This visit to see the baby Jesus was in all likelihood just one stop of many for the Magi in a day when travelers made slow progress and risked the hazards of bad weather, bandits, and sickness. This, of course would only make sense if the Magi were Persians as has been commonly assumed. However, if the Magi were Jewish priests from the East, the purpose for their visit would have been most likely wholly divine.

As we close this chapter on the fireball of the Magi we must ask one final question. If it was a meteorite, what ever became of it?" This question cannot be answered because the Bible is silent on the matter. The Magi are never mentioned again, and their star, likewise, suffers the same fate. One can only speculate, but the speculation is indeed tantalizing. One possibility is that the Magi took the stone with them. Although, admittedly, there is no evidence to suggest that they did. A few other possibilities exist that the stone might have been the revered stone of Elagabalus, or the black stone of the Kaaba. Once again, there is no evidence for these assertions, but both stones have enigmatic origins that seem to date from the early Christian era. The Roman historian Herodian writing in the 3rd century A.D. is the first to mention the stone of Elagabalus. The stone first appeared in history in a temple at Emesa, in Syria. The temple was dedicated to the pagan sun God, Elagabalus which was decorated in gold and silver. The object of veneration was a black stone that was said to be formed in a conical shape. The Syrians claimed that the stone had been hurled down to them from heaven by Zeus.

The stone gained some notoriety in the 3rd century when the Roman emperor Elagabalus (who had changed his name in honor of that Phoenician deity) apparently had the stone removed from the temple and brought to Rome. It was supposed to represent the Sun God. Elagabalus who was a sort of megalomaniac worshipped it much to the disgust of the Roman elite. Elagabalus was an eccentric, somewhat effeminate emperor who it was said, engaged in orgies and other perverted activities. Eventually a coup was raised against him and he was murdered, his desecrated body unceremoniously dragged to the sewer and discarded as if it were common trash. After his death the mysterious stone disappears from the records.

The black stone of the Kaaba is another object which is cloaked in mystery. This stone is located at the Masjid-al-Haram mosque in the Islamic holy city of Mecca in Saudi Arabia. It is unknown how the stone was first vaulted to its current status as one of Islams most revered relics. According to legend, the stone dropped from the heavens during the time of Adam and Eve, and Abraham was supposed to have built a temple at the site. At the time, it was said that the stone was pure white, but has blackened over the years due to the sins of mankind. It is said that Mohammed himself set the black stone in the corner of the mosque after settling a quarrel among some of the tribal factions in Mecca. The British explorer Sir Richard Burton visited the mosque in 1853 and left a vivid description of the relic. He wrote:

> The colour appeared to be black and metallic, and the centre of the stone was sunk about two inches below the metallic circle. Round the sides was a reddish brown cement, almost level with the metal, and sloping down to the middle of the stone. The band is now a massive arch of gold or silver gilt. I found the aperature in which the stone is, one span and three fingers broad.

It has long been thought that the black stone of the Kaaba is a meteorite. This, however, cannot be proven because of the stones inaccessibility. It is so revered by Muslims that for the time being anyway it is not available for scientific study. Like the stone of Elagabalus, the Kaaba stones origin seems to suggest that it has been around at least since the time of Jesus. Although there is no hard evidence that backs this up. It is only the legend of Abraham's temple built twenty centuries before Jesus that holds this theory up. However, there is also a story that the angel Gabriel had something to do with the stones arrival at the Kaaba. Could Gabriel merely be a metaphor for the fireball that brought the stone to Earth? Is it possible that Gabriel and the angel that visited the shepherds in the book of Luke are one and the same? Once again we are left to wonder.

Maximus Tyrius writing in the 2nd century mentions a stone that he describes as being "quadrangular." The Arabs worshipped this stone which represented some unknown god. This, of course, was five centuries before the birth of Islam. It is quite possible that this stone is the Kaaba stone, or perhaps the stone of the Magi. Lacking concrete evidence we possess only these tantalizing fragments at our disposal which we can attempt to piece together and create something that connects and is linear. At the moment, however, a speculative hypothesis is all that we can manage.

The Star of the Magi continues to fascinate people two thousand years after its appearance. To the secular audience it might seem that an analysis of it is hardly worth attempting. However, historians have attempted to make life out of subjects with far less credence than the Star of the Magi. One only has to look at all of the books written on the Holy Grail. This relic is not even mentioned in the Bible, at least under that name. The first mention of it seems to have occurred during medieval times by Geoffrey of Monmouth. Where Geoffrey obtained his information is a mystery but it seems plausible that he acquired it from a now lost Celtic source. Shortly after Geoffrey's work surfaced, a whole new genre was created, and the legend seems to have been born. Famous works written in the 13th century include the anonymous *Queste del Saint Graal* and Chrétien de Troyes *Story of the Grail*. These famous works have been followed over the intervening years by other works that have given the grail legend a new meaning. Some of these interpretations, although interesting, and fascinating to ponder, are hardly credible as far as the facts of the matter are concerned. The reason is that they all derive from the primary source of Geoffrey of Monmouth who was writing eleven centuries after the time of Jesus.

My point here is not to discredit the work done by these scholars whose efforts are at least done in sincerity and good faith, but to show that my interpretation of the Magi's star is at least as credible, and is based on evidence whose authors lived in close proximity to the times that the event occurred. As of today this cosmic mystery remains unsolved and open to further interpretation. I only hope that my interpretation has not muddied the waters too much. Only time will tell.

Chapter 15

THE DRAGON FIREBALL

For thousands of years the dragon has been a motif of good or evil in virtually every culture across the globe. The origin of the dragon is as mysterious as the image of the creature itself. In some countries the dragon was thought to have been real, and indeed still remains in that capacity to some. It existed in caves, or moats around a castle, and was sometimes seen emerging from the watery depths of some lake or ocean. At night, it might sit on a lonely precipice of some towering cliff keeping a watchful vigil on the town below. Indeed, the dragon is ubiquitous. No one is certain where it first originated, but there are plenty of clues left around the world that might lead us to its cosmic emergence.

An abundance of dragon myths exhibit certain characteristics which might show that the legends had a common origin, and then disseminated to other parts of the globe. Plenty of books have been written on the subject of dragons. The dragon has pervaded the civilization of man since the beginning of recorded history, and does not seem to want to go away. Why is this? What is it about the dragon that makes it timeless? This last question is the easiest to answer . . . The dragon is a nothing more than a metaphor. It is the stuff of life. In essence, it is the fireball that created us.

When one thinks of a dragon the image that comes to mind is usually of a fire breathing reptile that flies over castles, and wreaks havoc against a panic stricken and helpless populace, usually challenged by some stalwart hero who takes on the hellish beast and saves mankind. Sometimes, however, the dragon is a benign creature that merely keeps watch at the gates of a castle, or as Chinese mythology tells us, guards the mansions of the gods as T'ien Lung or the Celestial Dragon.

We have seen in earlier chapters of this book other examples of people interpreting a fiery cosmic body with something divine. A good example would be the stone that allegedly fell from the bosom of Cybele which ultimately became the personification of that deity. Since ancient times writers have likened fireballs and deep sky objects with dragons or other beastly creatures. Different cultures believed that the dragon possessed great powers. Indeed, the dragon is one of the twelve animals on the Chinese zodiac which represent the various segments of the sky. In Chinese folklore the dragon is known as "The spirit of the water" and possess' many traits that are normally associated with gods. For instance, the dragon is responsible for making the wind, and the course of rivers. In Japan, a great serpent is the keeper of the wells. The Mayans believed that serpents were needed to seed the earth and were able to control the rain. In Ohio there is a great mound shaped like a serpent or a dragon, obviously at one time it was used to venerate that creature. These myths from around the world offer an interesting speculation. It seems that the dragon is intertwined with the "stuff of life." Water, as we all know is the essential ingredient that is required for life to exist. Historically, most cultures in their ancient past believed in a form of animism. Polytheism evolved out of this and became prevalent during the 1st and 2nd millenniums B.C. This can be seen in the cultures of ancient Egypt, Sumer, classical Greece with its pantheon of gods and even Rome in its glory days. The dragon thrived during these times. With the advent of Buddhism, Christianity, and then Islam, this polytheism evolved into the monotheism and secular beliefs that we see today. The dragon, however, has survived it all. The fire breathing monster has weathered the times and remains with us, stubbornly refusing to go away. What does it all mean? Over the years people have formed their own conclusions. For the most part, the dragon denotes the dualistic forces of good and evil. At least this is one of the more popular sentiments. In world literature we see many cases where a great warrior takes up the challenge and whether the motive is mercenary, ideological, or merely out of love, he has combat with the dragon.

A good example of this is seen in the old Anglo-Saxon legend of Beowulf. Beowulf is a great warrior, king of the Geats who finds himself in the land of the Danes fighting with the cannibalistic monster Grendal who has been ravaging the court of the Danish King Hrothgar. Beowulf ends up wounding the monster and chases him back to the swamp where the creature resides with his mother. Worn out from the combat, Grendal makes it home only to die in front of his enraged mother who seeks vengeance in his name. She goes on a murderous rampage killing one of Hrothgar's ablest and loyal officers and devouring his body. Beowulf and a small contingent of warriors go in pursuit, and follow the beasts tracks back to the swamp where she has retreated to rest. Beowulf alone enters the monsters abode and engages Grendal's mother in combat. After a brutal fight she is slain and Beowulf emerges victorious from the swamp carrying Grendal's head which he presents to the awestruck dignitaries of the Danish court.

After these exploits, Beowulf is immortalized. He returns to his land and lives for years in peace. However, at an advanced age, he once again finds himself called upon to perform a heroic deed. This time the combat is with a mighty dragon. According to the legend, this dragon kept guard over a horde of gold and other treasure that is located in the mouth of the dragon's cave. One day, when the great serpent is asleep, a bold warrior who has heard rumor of this wealth sneaks in and is able to make off with a golden cup. The dragon awakens to find that someone has trespassed in its lair and it is not long before it discovers the missing cup. The beast then works itself into a frenzy and ravages and terrorizes the countryside, breathing fire and stench on Beowulf's kingdom.

Beowulf is called to action, and with a hand picked group of elite warriors sets out to confront the monster. The fearless dragon is not hard to find and the aging Beowulf challenges it to combat. Beowulf, however, is dismayed to find that not one of his warriors has the courage to face the beast. He therefore is forced to go at it alone. Unfortunately, he is no longer the fierce and formidable warrior that he once was, though he still possess' that indomitable spirit of his youth. His body is aged and rusty, but he holds his own with the dragon until his sword breaks at the hilt. The dragon then delivers Beowulf a crushing blow that disables him. Just as it appears that the great hero is about to be slain, one of his warriors named Wiglaf has somehow mastered the courage to assist his aged king. Beowulf, buoyed by this new turn of events manages to regain his feet and together the two men are able to slay the dragon. For Beowulf, however, his wounds have proven more than his body can endure and he dies a warrior's death.

Another famous encounter with a dragon occurs in the legend of St. George. Not much is known of the real man from whom the legend was born. Apparently he lived during the late 3rd or early 4th century A.D. and was a champion of the Christian cause while serving as an officer In the Roman army under the emperor Diocletian. It was said that he was martyred in either 304 or 305 A.D. It was after his death that St. George became immortalized.

It is known that he hailed from Jaffa in Palestine, and it is from that place where the legendary encounter with the dragon takes place. In 1230 A.D. Jacobus de Voragine, the Bishop of Genoa wrote a work entitled *The Golden Legend*. This work, which has now been ingrained into the foundation of the Christian faith tells the most popular version of the legend. Near Lydda, there was a small pond in which a dragon resided. Every so often he required an offering from the town as payment for not destroying it in full. At first the offering was made with sheep, but the dragon soon tired of this and decided to up the price. He decided that he would accept nothing but a human sacrifice. The townspeople were, of course appalled at this demand but they had no choice but to submit to the dragons will. They therefore devised a method of who would should be sacrificed by drawing lots. Rich or poor, it did not matter, no one was exempt from taking part.

It came to pass one day that the king's daughter had drawn the dreaded lot. The king was forced to make a tough decision; lose the faith and trust of his people, or give his daughter up. He sorrowfully chose the latter and sent his poor daughter from the gates of the city to the dragon's lair. However, along the way, it just happened to be her good fortune to cross paths with St. George. The knight asked the young girl where she was headed in such a melancholy state. She broke down and told St. George all about the dragon and then begged him to flee lest he die with her. St. George, however, was a true knight bound to the code of chivalry. Fleeing from danger, especially when a damsel needed his services was not an option that he could adhere to. He told the king's daughter to fear not, for as long as he lived no harm would come to her from any dragon.

George then confronted the dragon and pierced it with his lance. St. George decided to tame the wounded beast and take it back to the town to show the people that they no longer had anything to fear. The townspeople were elated by this new turn of events and St. George seized the moment and introduced Christianity to the people who readily accepted this new creed. This is merely one of many legends that surround St. George. Over the centuries there have been many purported sightings of this saint. He is the patron saint of soldiers and has supposedly been seen at numerous battles leading men into the heart of the action. He was said to have been present with Richard the Lionheart at the siege of Acre during the 3rd crusade in 1190. However, probably the most famous sighting took place during the 1st World War when he was supposed to have been present with British troops at the Battle of Mons. It was even said that he led a cavalry charge mounted on a white stallion during the course of that battle.

Illustration 17

Leaving Beowulf and St. George we turn now to a Japanese legend which was supposed to have taken place during the 10th century. A young girl named Ki Yohime fell in love with a monk named Anchin. The monk, however, was too caught in the vows of chastity to be able to respond to the girl's passion. In an attempt to thwart her advances he hid under a large bell at the Dojiji temple in Kyoto. However, the girl was undaunted by the monks rebuke and conjured up the spirit of a dragon into which form she changed. The terrified monk could do nothing but watch as the dragon wrapped itself around the bell. Its great weight caused the bell to collapse pinning the unfortunate monk underneath it. The dragon then breathed a fire so hot that it caused the bell to melt which killed the monk as well as his stalker.

Although the fire breathing dragon is a fiction It was believed by many to really exist in this form until recent times. In 1567 the Cambridge scholar John Maplet wrote about an encounter between a dragon and an elephant which he included in his often entertaining work of natural history, *A Greene Forest:*

> The dragon is the heade and chiefest of all other serpents, and
> flieth from his den or cave in the earth his holownesse up to the
> top of the brode ayre, and of Dragon in Greek, is englished flight.
> Plinie saith, that betweene the dragon and the elephant there is a
> naturall warre. Insomuch as the dragon enrowleth and twineth
> about the elephant with his taile, and the elephant againe with
> his snoute used as his hande, supplanteth and beareth downe
> the dragon: The dragon with twining about him holdeth fast,
> and with his might somewhat bendeth backwards the head and
> shoulders of the elephant, which being so sore grieved with such
> waight, falleth downe to the ground first, and is therewithall
> slaine: but that other scapeth not scotfree, for with one anothers
> holde and rushing to the ground the elephant also is broused,
> and after withall slain. Againe they strive togither after this sort.
> The elephant espying him sitting on the loft of a tree, runneth
> as fast as he can with full brut to that tree, hoping thereby to
> shake downe the dragon, and to give him a deadly fall: but

in that he doth not after the wysest sort for him selfe. For the dragon so falling, oftentimes lighteth on his necke or shoulders, and agrieveth him as with byting at his nostrelles, and pecking at his eies, and sometime he dazeleth him, and goeth behinde at his back and sucketh out of his bloud, so that if he shaketh him not off betimes by sucke wasting of bloud as he will make, thereby he is quickly enfeebled: he falleth downe heavily with the dragon also holding aboute him, and are killed both with so heavie and burdenous a fall.

As can be inferred from reading the above description, Maplet, it is obvious, never observed a dragon first hand, but had implicit faith that the creature existed. He seems to have derived his information from other sources, namely Pliny, a source that was 1500 years old even in Maplet's time. The above account seems ludicrous. It is possible that the serpent like animal described is a large snake of some kind, but it is hard to imagine the two creatures grappling in the manner described. It is known that in England and other European countries during these times, that exotic animals were sometimes used in a sort of gladiatorial capacity for entertainment purposes. Perhaps Maplet is unknowingly alluding to one of these instances. Whatever the case, this example shows that if a person as erudite as Maplet believed that dragon's existed, what must a person of average or below average intelligence have thought of them?

Times have changed since Maplet's day. People no longer believe in the dragon in the form in which he believed it to exist. A lot of this has to do with the shrinking size of our planet. Maplet lived during the age of discovery. The Eastern and Western hemispheres were just getting acquainted with one another. Explorers like Balboa and Magellan set off from their home ports with only a vague idea of where they were going. The maps that they had in their possession were crude and hypothetical. Largely the result of guesswork and hearsay. Interesting enough, some of these early attempts at cartography showed not only spurious land masses, but sea monsters of all sorts including creatures that looked remarkably like dragons.

Over the course of the last few centuries the world would become smaller as advances in engineering and nautical skills improved. Maps became more accurate so that explorers, traders, and global speculators now had a clearer picture of where they were going and how long it might take to get there. Dragons slowly descended into the realm of folklore and legend. No longer did the dragon strike fear into the hearts of people who looked skyward and saw a ball of fire streaking across the night sky. However, did our ancestors really believe that fireballs were caused by the venomous fire of a dragon? You bet they did! And the evidence is in the chronicles in which they left behind.

In the year 1203 the Chronicles of the Mayors of London reported that "In London there fell great raines, thunderings, and hales (stones as big as eggs) whereby many trees and corne were destroyed; and birds were seen flying in the "ayre" with "fyre" in their mouthes, and to set fyre in houses & burn them."

In 793, over England dragons were seen in the sky shooting fire. In 1222 two reports from two different sources tell us the same thing. The Waverly Annals tell us that dragons were seen flying together and some witnesses said that they were even fighting. William of Newburgh reports that a comet was seen accompanied by dragons.

The reports just mentioned show that the dragon was believed to be an actual fire-breathing bird-like creature in the traditional fairy tale sense for at least a millennium.

However, the reality of the fierce fire-breathing dragon seemed to wane out as the 18th century and the "Age of Reason" hit full stride. In 1738 it is recorded that a great meteor was seen in Dorsetshire the kind that naturalists call the "Draco Volans" or "The flying dragon." By then, anyway, it seems that the dragon was looked at in a metaphorical way.

Although the dragon no longer threatens us with its flaming tongue, it nevertheless remains a potent symbol of what and who we are. Today, when we think of the dragon we think of its violent nature that our ancestors handed down to us in the form of myth and folklore. They are monstrous looking reptiles with a long spiny tails, scaly green bodies and fierce looking heads that belch fire from a grotesque mouth that emanates an intolerable stench.

Are dragons nothing but a mere symbol of our origins? The Aztecs were once the architects of a vast empire that flourished in Mexico up until the time of the Spanish invasion in the early 16th century. Their chief god was Quetzalcoatl who is often seen as a large eagle holding a serpent. Indeed, today this image is seen on the flag of Mexico. Quetzalcoatl was a god that reigned above all others, and the Aztecs paid homage to him usually in the form of sacrifice. This most often took the form of a bull or some other type of animal, but oftentimes the victim was human, usually a captive taken from battle. Quetzalcoatl's father was Citallatonali who lived in the sky and was the dragon from which the Earth was made. This here is a stunning revelation and perhaps gives us a clue of what the dragon actually represented in its incipient form. Why did the Aztecs associate dragons with the origin of our planet? Amazingly, evidence shows that they were not the only ones.

In Greek mythology the Gorgons were serpent-like creatures with wings. They were immortal and it was said that if they looked at someone, that individual would turn to stone. Is this perhaps a subtle reference to a meteorite? There are countless other stories that associate the dragon with the origin of our planet. Why is this? What is going on here? If the dragon is a metaphor of a fireball, it would seem that the message that our ancestors left us was subliminal. Our planet formed over 4 billion years ago. However, if we were to travel back to this period of Earth's antiquity we would quickly notice that it is not the same place that we know so well today. In fact, not only could a person not stand on its molten surface but since there was no atmosphere a person could not breathe as well. However, just for kicks, let us say that you could stand on its scolding surface. Perhaps on one of the first rock outcrops to form on our planet. What do you think that you would see if you glanced skyward? Barring the sulfuric clouds you would by all accounts be a witness to the most spectacular and most enduring meteor shower to ever to hit the Earth. The sky would be full of fireballs raining down onto the surface of our hellish looking world. You would in essence be viewing the birth of our planet. Is this the subliminal message that the dragon conveys to us? Was primitive man somehow privy to information regarding the violent origin of our planet? This hardly seems possible. But is it somewhere, somehow indelibly impressed into the fabric of our being?

Obviously, in these early years of our planet's history, Earth was inhospitable to any form of life no matter how primitive. After all, the accretionary disc of our solar system had only recently formed. The Earth was being bombarded by comets, meteors and other stellar bodies on a regular basis, therefore giving it no chance for the necessary ingredients of life to evolve. In fact, it is now believed that a massive proto planet slammed into the Earth early in its history. This body, upon hitting the surface would have exploded and left a ring of molten rock revolving around the Earth. Over time this rock would have amalgamated into one body which we now know as the moon.

It would take hundreds of millions of years, but eventually the constant bombardment of cosmic bodies would taper off to a point where there was now a period of relative stability. The Earth's surface managed to cool down and the oceans were able to form due to just the right amounts of the lighter elements present in our early atmosphere. After the formation of the oceans it was only a matter of time before the first life showed itself in the form of Prokaryotic cells. Is this the reason that the dragon is commonly and almost universally associated with the ocean?

The only thing left to be said about the dragon metaphor has to do with the duality of its nature. Good or bad the dragon symbolizes the struggle of the cosmic elements. When dealing with life, the survival of each organism, no matter how simple its structure is dependent on the actions of those organisms that it is forced to cohabit with. A series of events that started with a ball of fire billions of years ago has led to the world that we see today. Beowulf, St. George, and Kiyohime are merely a few of the subjects who carry on the internal struggle of this subliminal nature. They are the messengers of our birth. They cannot help it, for it is the nature of all things. It is the symbiotic relationship that occurs with everything. In essence we are all one. Good or bad, we are the same. We are the dragon . . . The fireball from which all things were created.

Bibliography (Books)

Allies, Jabez. *On the Ancient British, Roman, and Saxon Antiquities and Folk-Lore of Worcestershire.* London, J.H. Parker, 1852.

Archbold, Rick. *Reliving the Era of Great Airships: Hindenburg an Illustrated History.* Toronto, Madison Press Books, 1994.

Baring-Gould, S. *Lives of the Saints.* London, John Hodges, 1882.

Baxter, John & Atkins, Thomas. *The Fire Came By: The Riddle of the Great Siberian Explosion.* New York, Doubleday & company, 1976.

Bede. *The Ecclesiastical History of the English People.* London, Penguin Books, 1990. (6)

Bergreen, Laurence. *Over the Edge of the World: Magellan's Terrifying Circumnavigation of the Globe.* New York, Harper Collins, 2004.

Blake, John F. *Astronomical Myths: Based on Flammarion's "History of the Heavens."* London, Macmillan & Co., 1877.

Bostock, John & Riley H.T. *The Natural History of Pliny: Translated With copious Notes and Illustrations.* London, George Bell & Sons, 1893. (3)

Bowditch, Henry Ingersoll. *Memoir of Nathaniel Bowditch.* Boston, James Munroe & Company, 1841.

Braddon, M.E. *Belgravia: A London Magazine.* London, 1868.

Bradford, William. *Mourt's Relation.*

Brand, John. *Observations on Popular Antiquities.* London, Charles Knight & Co., 1842.

Brewster, David. *The Edinburgh Encyclopedia: Chronological History of Meteorites, Interspersed With Remarks.* Edinburgh, William Blackwood, 1830.

Bridgeford, Andrew. *1066: The Hidden History in the Bayeux Tapestry.* New York, Walker & Company, 2006. (9)

Britton, C.E. *A Meteorological Chronology To A.D. 1450.* London, Meteorological Stationary Office, 1937. (4)

Brocklesby, John. *Elements of Meteorology: With Questions for Examination, Designed for Schools and Academies.* New York, Sheldon and Company, 1869.

Burke, John G. *Cosmic Debris: Meteorites in History.* Berkeley and Los Angeles, University of California Press, 1986.

Butler, Samuel. *The Odyssey of Homer.* New York, Classics Club, 1944.

Campbell, Joseph. *The Hero with a Thousand Faces.* Princeton, Princeton University Press, 1973.

Carus, Paul. *Chinese Thought: An Exposition of the Main Characteristic Features of the Chinese World-Conception.* Chicago, The Open Court Publishing Company, 1907.

Cavendish, Cecil. *Will-O'-the Wisp in St. Nicholas: An Illustrated Magazine for young Folks.* New York, The Century Company, 1910.

Cerveny, Randy. *Freaks of the Storm: The World's Strangest True Weather Stories.* New York, Thunder's Mouth Press, 2006.

Comte, Fernand. *Mythology.* Edinburgh, W & R Chambers, 1991.

Cotterell, Arthur. *The Penguin Encyclopedia of Ancient Civilizations.* London, Penguin Books, 1988.

Croker, Crofton T. *Fairy Legends and Traditions of the South of Ireland.* London, John Murray, 1828.

Cumont, Franz. *The Oriental Religions in Roman Paganism.* Chicago, The Open Court Publishing Company, 1911.

Darrach, H.B. & Ginna, Robert. *Have We Visitors from Space?* In Life Magazine, April 7, 1952.

Day, Malcom. *A Treasury of Saints:100 Saints: Their Lives and Times.* New York, Barron's, 2002.

De Fonvielle, W. *Thunder and Lightning.* New York, Scribner and Armstrong,

Dersin, Denise. (Editor) *What Life Was Like: In the Lands of the Prophet: Islamic World AD 570-1405.* Richmond Virginia, Time-Life Books, 1999.

Doniger, Wendy. *The Rig Veda.* London, Penguin Books, 1981.

Dor-Ner, Zvi, *Columbus: The Age of Discovery.* New York, William Morrow and Company, 1992.

Dyer, T. F. Thiselton. *The Will-O'-The Wisp and its Folk-Lore in Gentleman's Magazine.* London, Chatto & Windus, 1881.

Echols, Edward C. *Herodian of Antioch's History of the Roman Empire From the Death of Marcus Aurelius to the Accession of Gordian III.* Berkeley & Los Angeles, University of California Press, 1961.

Edersheim, Alfred. *The Life and Times of Jesus the Messiah.* London, Longmans Green, 1883.

Edwards, Frank. *Flying Saucers-Here and Now.* New York, Lyle Stuart, 1967.

Einhard And Notker The Stammerer. *Two Lives of Charlemagne.* London, Penguin Books, 1969. (7)

Flammarion, Camille. *Thunder and Lightning.* London, Chatto & Windus, 1905.

Fletcher, L. *An Introduction to the Study of Meteorites: With a List of the Meteorites Represented in the Collection.* London, William Clowes and Sons, 1896.

Fort, Charles. *The Book of the Damned: The Collected Works of Charles Fort.* New York, Tarcher/Penguin, 2008. (16)

Frankland, E. & Lockyer, J.N. *A Dictionary of Science, Literature & Art. Volume II Page 608.* London, Longmans, Green & Co., 1866.

Furneaux, Rupert. *The Tungus Event: The Unsolved Mystery of the World's Greatest Explosion.* Great Britain, Panther Books, 1977.

Gallant, Roy A. *The Day the Sky Split Apart: Investigating a Cosmic Mystery.* New York, Atheneum, 1995.

Geoffrey of Monmouth. *The History of the Kings of Britain.* Great Britain, Penguin Books, 1982.

Gervase of Canterbury. *The Chronicle of the Reigns of Stephen, Henry II, and Richard I.* London, Longman & Co. 1879. (10)

Giles, J. A. *Matthew Paris's English History: From the Year 1235 to 1273.* London,

George Bell & Sons, 1889.

Gordon, E. O. *Saint George: Champion of Christendom and Patron Saint of England.* London, Swan Sonnenschein & Company, 1907.

Gorky, Maxim. *Untimely Thoughts: Essays on Revolution, Culture and the Bolsheviks 1917-1918*. New Haven & London, Yale University Press, 1995.

Grant, Michael. *Myths of the Greeks and Romans*. New York, Mentor, 1986.

Gregory Of Tours. *The History of the Franks*. London, Penguin Books, 1974. (5)

Griffin, S.G. *A History of the Town of Keene*. Keene N.H. Sentinel Printing Company, 1904.

Guerber, H.A. *Myths of the Norsemen: From the Eddas and Sagas*. New York, Dover Publications, 1992.

Hamilton, Edith. *Mythology*. Boston, Back Bay Books, 1998.

Harrison, Jane Ellen. *Prolegomena to the Study of Greek Religion*. Cambridge, University Press, 1903.

Hartwig, Gustav. *The Aerial World: A Popular Account of the Phenomena and Life of the Atmosphere*. New York, D. Appleton and Co. 1875.

Hastings, James. *Dictionary of the Apostolic Church*. Edinburgh, T.&T. Clark, 1918.

Herodotus. *The Histories*. New York, Barnes & Noble Classics, 2004.

Houston, Edwin J. *The Wonder Book of the Atmosphere*. New York, Frederick A. Stokes Company, 1907.

Hughes, Thomas Patrick. *Dictionary of Islam*. London, W.H. Allen & Co., 1885.

Huntington, Oliver Whipple. *Popular Science Monthly: A Talk About Meteorites*. New York, D. Appleton And Company, July, 1890.

Ingersoll, Ernest. *Dragons and Dragon Lore*. New York, Dover Publications, 2005.

Irving, Washington. *The Life and Voyages of Christopher Columbus*. Philadelphia, David McKay Publisher, 1893. (12)

James, Barnard J. *The Living Age: Nature's Night Lights*. Boston, The Living Age Company, 1911.

Jastrow, Robert. *Red Giants and White Dwarfs: The Evolution of Stars, Planets and Life*. New York, Harper & Row, 1967.

Kinsella, Thomas. *The Tain: From the Irish Epic Tain Bo Cuailnge*. New York, Oxford University Press, 2002.

Kirkwood, Daniel. *Meteoric Astronomy: Shooting Stars, Fire Balls, and Aerolites.* Philadelphia, J.B. Lippincott & Co. 1873. (1)

Krinov, E.L. *Giant Meteorites.* New York, Pergamon Press, 1966.

Lackland, William. *Meteors, Aerolites, Storms, and Atmospheric Phenomena.* New York, D. Appleton & Co. 1870.

LaPaz, Lincoln and LaPaz Jean. *Space Nomads: Meteorites in Sky, Field, and Laboratory.* New York, Holiday House, 1961.

Leslie, Desmond. & Adamski, George. *Flying Saucers Have Landed.* New York, The British Book Centre, 1955.

Livy. *A History of Rome: Summaries, Fragments, Julius Obsequens,* Cambridge, Harvard University Press, 2004. (2)

Lowe, E.J. *Natural Phenomena and Chronology of the Seasons.* London, Bell & Daldy, 1870. (11)

Maplet, John. *A Greene Forest.* The Hesperides Press, London, 1930.

Marshak, Stephen. *Earth: Portrait of a Planet.* New York, Norton and Company, 2005.

Matarasso, P.M. *The Quest of the Holy Grail.* London, Penguin Books, 1969.

Mather, Increase. *Remarkable Providences: Illustrative of the Earlier Days of American Colonisation.* London, Reeves and Turner, 1890. (15)

Maxwell-Stuart, P.G. *Chronicle of the Popes: The Reign-by-Reign Record of the Papacy From St. Peter to the Present.* London, Thames and Hudson, 1997.

McCarthy, Justin. *Irish Literature: Will o' the Wisp From Hibernian Tales, A Chap Book.* Philadelphia, John D. Morris & Co., 1904.

Muir, J. *Original Sanskrit Texts of the Origin and History of the People of India, Their Religion and Institutions.* London. 1883.

Nicholson, William. *The Journal of Natural Philosophy, Chemistry & the Arts: Volume XXVIII.* London, Stratford, 1811.

Paine, Thomas. *Common Sense: The Life and Works of Thomas Paine Volume II.* New York, Thomas Paine Historical Association, 1925.

Pare, Ambrose. *On Monsters and Marvels: Translated by Janis L. Pallister.* Chicago and London, The University of Chicago Press, 1983.

Parnell, Arthur. *The Action of Lightning: and The Means of Defending Life and Property from its Effects.* London, Crosby Lockwood and Company, 1882.

Phipson, T. L. *Meteors, Aerolites, and Falling Stars*. London, Lovell Reeve & Co. 1867.

Plutarch's Lives. Volume's I & II. New York, The Modern Library, 2001.

Ruppelt, Edward J. *The Report on Unidentified Flying Objects*. New York, Ace Books, 1956

Sagan, Carl. And Page, Thornton. *UFO's A Scientific Debate*. New York, Norton, 1974.

Sandars, N.K. *The Epic of Gilgamesh*. London, Penguin Books, 1972.

Scarre, Chris. *Chronicle of the Roman Emperors: The Reign-by-Reign Record of the Rulers of Imperial Rome*. New York, Thames and Hudson, 1995.

Scarre, Chris. *The Penguin Historical Atlas of Ancient Rome*. London, Penguin Books, 1995.

Short, Thomas. *A General History of the Air, Weather, Seasons, Meteors in Sundry Places and Different Times; More Particularly for the Space of 250 Years: Together with Some of Their Most Remarkable Effects on Animal (Especially Human) Bodies, and Vegetables*. London, T. Longman.

Silliman, B. & Kingsley, J.L. *The Monthly Anthology and Boston Review: Containing Sketches & Reports of Philosophy, Religion, History, Arts and Manners*. Boston,

Snelling & Simons, 1808.

Snowe, Joseph. *The Rhine: Legends, Traditions, History, from Cologne to Mainz*. London, J. Madden & Co. 1839.

Spence, Lewis. *History of Atlantis*. London, Senate, 1995.

Slafter, Carlos. *Sir Humfrey Gylberte and His Enterprise of Colonization in America*. Boston, Prince Society, 1903.

Strauss, David Friedrich. *The Life of Jesus: Critically Examined*. London, Swan

Sonnenschein & Company, 1892.

Suetonius. *The Twelve Caesars*. London, Penguin Books, 1989.

Temple, Robert K.G. *The Sirius Mystery*. Rochester, Vermont, Destiny Books, 1987.

Terry, Milton S. *The Sibylline Oracles: Translated From the Greek Into English Blank Verse*. New York, Hunt & Eaton, 1890.

Vaeth, Gordon J. *200 Miles Up: The Conquest of the Upper Air.* New York, The Ronald Press Company, 1951.

Vallee, Jacques. *Anatomy of a Phenomenon: UFO's in Space.* New York, Ballantine Books, 1974.

Velikovsky, Immanuel. *Worlds in Collision.* Garden City, N.Y. Doubleday & Company, Inc. (17)

Verma, Surendra. *The Mystery of the Tunguska Fireball.* Cambridge, Icon Books, 2006.

Werner, E.T.C. *Myths and Legends of China.* New York, Dover Publications, 1994.

White, Andrew D. *A History of the Doctrine of Comets.* New York & London, G.P. Putnam's Sons, 1887.

Wilson, H.H. *Rig-Veda Sanhita: A Collection of Ancient Hindu Hymns Constituting the Fifth Ashtaka, or Book, of the Rig-Veda; The Oldest Authority for the Religious and Social Institutions of the Hindus.* London, N. Trubner and Co., 1866.

Winchester, Simon. *Krakatoa: The Day the World Exploded August 27, 1883.* New York, HarperCollins, 2003.

Winsor, Justin. *Christopher Columbus: And How he Received and Imparted the Spirit of Discovery.* Stanford CT, Longmeadow Press, 1991. (13)

Yeomans, Donald K. *Comets: A Chronological History of Observation, Science, Myth, and Folklore.* New York, John Wiley & Sons, Inc.

OTHER SOURCES

"A Possible Impact Crater for the 1908 Tunguska Event." Blackwell Publishing 2007. www.blackwell-synergy.com

"Ball Lightning: A Shocking Scientific Mystery." National Geographic on the Web, 31 May, 2006. <http://news.nationalgeographic.com/news/pf/32051113.html>

"Ball Lightning Scientists Remain in the Dark." 20 December, 2001. <http://www.newsscientist.com/article.ns?id=dn1720>

"Benjamin Silliman." <http://www.peabody.yale.edu/archives/ypmbios/silliman.html>

Catalogue of Aerolites and Bolides from A.D. 2 to A.D. 1860—30th Meeting of the British Association, London, John Murray, 1861. (14)

Chamberlain, Von Del & Krause David J. "The Fireball of December 9, 1965-Part 1:

Calculation of the Trajectory and Orbit by Photographic Triangulation of the Train." The Royal Society of Canada.

"Crater Could Solve 1908 Meteor Mystery." MSNBC on the Web, 26 June, 2007. http://www.msnbc.msn.com/id/19436962/print/1/displaymode/1098/

"Daedalus & Icarus." <http://thanasis.com/icarus.htm>

"Fireball Fears Stoked by Space history." MSNBC on the Web. 29 March, 2007 <http://www.msnbc.msn.com/id/17859275/

"Fire in the Sky." Time on the Web. <http://www.time.com/time/magazine/article/0,9171,894885,00.html>

Holleman, Joey. "Did a Comet Destroy a Civilization and Kill the Mammoths 12,900 Years Ago? An S.C. Site Could Provide Evidence." The State. 7 October, 2007. (8)

"Hydrogen May Not Have Caused Hindenburg's Fiery End." New York Times on the Web, 6 May, 1997. <http:query.nytimes.com/>

"Mystery Space Blast 'Solved.'" BBC on the Web, 30 October, 2001. http://bbc.co.uk/1/hi/sci/tech/1628806.stm

Nennius: Historia Brittonum. Medieval Sourcebook. <http://www.fordham.edu/halsall/basis/nennius-full.html>

"New Economic Policy." <http://en.wikipedia.org/w/index.php?title=New_Economic_Policy>

"People in Kecksburg want to resolve what fell from the sky in 1965." Pittsburgh Post-Gazette on the Web, 09 March, 2003. http://www.post-gazette.com

"Project Twinkle Final Report." <http://www.project1947.com/gfb/twinklereport.htm>

"Report of the Thirtieth Meeting of the British Association for the Advancement of Science; Held at Oxford in June and July 1860. London, John Murray, 1861.

"Space Fireballs Sighted from Jetliner." MSNBC on the Web, 29 March, 2007. <http://www.msnbc.msn.com/id/17836220/>

"The Catholic Encyclopedia: An International Work of Reference on the Constitution, Doctrine, Discipline, and History of the Catholic Church." New York, The Encyclopedia Press, 1913.

"The London Encyclopedia: or Universal Dictionary of Science, Art, Literature, and Practical Mechanics. London, Thomas Tegg, 1829.

"The Mystery of the Green Fireballs." Burbank, California, W.L. Moore Publications. 1989.

"The Thunderstorm: The Nature, Properties, Dangers, and Uses of Lightning. London, 1848.

"UFOs in History." http://www.bibleufo.com/ufos.htm

Washburn, Mark. "What Was That Light in the Sky." The Charlotte Observer. 26 January, 2007.

"What Was That? Strange Lights in Upstate Skies." WYFF4 on the Web, 26 January, 2007. http://www.wyff4.com

World Heritage Center. http://whc.unesco.org

Atmospheric Phenomena Chronology

Pre-Recorded History

251 million years ago—This was known as "The Great Dying." 90% of aquatic species, and nearly 75% of land species disappeared from the Earth. The cause of this mass extinction is unknown, but it is generally thought to have been brought on by either a comet or a meteor impact.

65 million years ago—The age of the dinosaurs comes to an end. Recent science has brought to light a possible culprit being the impact of a large off the Yucatan peninsular near Mexico.

12,900 B.C.-Recent evidence has indicated that a comet may have exploded, or broke apart in the Earth's atmosphere leading to mass extinctions which included the mastodon and the mammoth.

Recorded History

1478 B.C.—The Parian Chronicle mentions a fireball, or a "thunderstone" which fell near or on the island of Crete. This fall may or may not have had something to do with the sudden disappearance of the Minoan civilization that had been flourishing up until this time. Other evidence points to the volcanic eruption of nearby Mt. Thera as the cause. (1)

1168 B.C.—A large rock or chunk of iron fell in the area around Mt. Ida on the island of Crete. (1)

687 B.C-(March 23) The Assyrian army of Sennacherib is annihilated by a "heavenly blast." Immanuel Velikovsky has determined from various sources, including Edouard Biot's Catalogue, that this occurred on March 23, 687 B.C. (17)

654 B.C.—According to Livy some stones fell on Alban Hill near Rome. (1)

616 B.C.—(January 04) The Chinese Annals states that a stone from heaven fell to Earth striking several chariots and killing 10 men. (1)

466 B.C.—A stone referred to by the ancient writers as "The mother of the gods" fell either near Thrace or Anatolia (modern day Turkey.) Legend has it that the poet Pindar was sitting on a hill in a contemplative mood when the meteorite slammed into the Earth near his feet. The stone was said to have been encircled by a fire as it descended through the sky. Upon striking the Earth, the meteorite was described as being "of moderate dimensions, of a black hue, of an irregular, angular shape, and of a metallic aspect." This stone is thought to have been the one that was brought to Rome in 204 B.C. by Scipio and used as an idol to worship the Phrygian goddess Cybele in a last ditch effort to find a way to defeat the Carthaginian general Hannibal. Within months Hannibal was indeed soundly defeated at the battle of Zama. (1)

190 B.C.—At Tusculum, southeast of Rome, there was said to have been "a shower of Earth." (2)

188 B.C.—On Aventine Hill, in Rome, stones were seen falling from the heavens. (2)

186 B.C.—At Picenum, in Italy, there was a shower of stones. (2)

167 B.C.—At Anagnia, in central Italy there was a shower of earth. (2)

167 B.C.—At Lanivium, near Rome, a blazing fireball was seen streaking across the sky. According to the Roman scribe Julius Obsequens, another

fireball was seen at Lanivium the following year. Since Obsequens was writing nearly 500 years after the event, and relying almost totally on Livy as his source, the two fireballs are in all likelihood one and the same. (2)

163 B.C.—At Cephallonia, in Greece, there was heard a sound like a trumpet in the sky followed by a shower of earth. (2)

162 B.C.—At Anagnia it was said that the sky seemed to be on fire at night. This could have been caused by the Aurora Borealis. However, it must be noted that it is rare to see this phenomena at this latitude. (2)

154 B.C.—At Compsa, in Italy, weapons were seen flying through the sky. This was almost certainly a meteor shower. (2)

152 B.C.—A stone shower was seen at Aricia (a suburb of Rome.) At this time a day of prayer was called for after strange things were seen in the sky. Some places reported seeing apparitions of men in togas which disappeared when people approached them. (2)

147 B.C.—A comet was seen in the sky for 32 days. (2)

140 B.C.—At some locations near Rome, images were seen falling from the sky. (2)

137 B.C.—Near Praeneste, in Italy, a fireball was seen in the sky. Thunder was also heard although there was not a cloud in the sky. (2)

133 B.C.—At Ardea, south of Rome, stones fell from the sky. (2)

125 B.C.—At Arpi, in Italy, it rained stones for 3 days. (2)

122 B.C—In Gaul (modern day France) 3 suns and 3 moons were seen in the sky. This sounds like parhelion, but might be something else entirely. (2)

113 B.C.—In Gaul, the sky caught fire. (2)

106 B.C.—A meteor was seen at Rome accompanied by a loud noise and javelins falling from the sky. (2)

104 B.C.—In the eastern and western sky weapons were seen battling. (2)

102 B.C.—At Etruria stones were seen falling from the sky. (2)

100 B.C.—At Tarquinia, in Italy, a blazing fireball was seen soaring toward the Earth. Also, at around sunset, a shield-like object was seen streaking across the sky from west to east. (2)

94 B.C.—Julius Obsequens mentions a meteor that appeared which seemed to set the whole sky on fire. Location unknown. (2)

93 B.C.—At the town of Volsinii in Italy, a flame issued forth from the sky at dawn. The sky opened up and split in two. In the opening flames appeared. (2)

91 B.C.—In Rome, as the Sun was rising a fireball blazed across the northern sky with great noise. (2)

87 B.C.—At Rome, the sky opened up and struck down many soldiers in the camp of Pompey along with weapons and standards. Julius Obsequens states that Pompey "perished by the blast of a heavenly body." Interesting enough, this incident was also mentioned by Pliny the Elder and Petronius. Pliny said that "Pompeius was paralyzed by a star." To diffuse any confusion the reader might be having, the Pompey mentioned here was not the same Pompey that was a third of Rome's First Triumvirate along with Julius Caesar and Crassus. Indeed, this man was his father. It is possible that Pompey was the unfortunate victim of a lightning strike. However, it is plausible that this could have been one of those rare instances of someone being struck by debris from outer space. (2) (3)

44 B.C.—During the festival of the Mother Venus, a comet appeared in the heavens. This occurred shortly after the assassination of Julius Caesar. The new emperor, Octavius (Caesar Augustus) dedicated this celestial object to the late Caesar. It was generally thought among the Roman populace that this comet was the vehicle that carried Caesar's soul to heaven. Also, at around this same time, a meteor was seen traveling through the western sky. (2)

42 B.C.—3 suns were seen in the sky over Rome that inevitably amalgamated into one as the day progressed. Many people believed that this was a representation of Rome's Second Triumvirate of Octavius, Antony, and Lepidus. A more scientific explanation would suggest that this was nothing more than parhelion (a sundog) which is a process that occurs when sunlight is reflected off of ice crystals in the atmosphere. (2)

17 B.C.—A meteor streaked across the Roman sky from south to north which made it seem as if night had turned into day. (2)

7 B.C.—A great fireball is seen somewhere to the east of Bethlehem in Judea. Most probably either in Persia, or in the region around the Dead Sea. Admittedly, this is speculative, but it seems that the New Testament books of Luke and Matthew allude to this. (Author)

30 A.D.—Eochie Oireaw, an Irish king is "slain and burnt from lightning fire from heaven." (4)

85 A.D.—Battles were seen in the air over Britain. (4)

112 A.D.—Thomas Short mentions a battle seen in the air over Britain. (4)

230 A.D.—Soldiers and horsemen were seen in the sky over London and other places in England. (4)

249 A.D.—There was a bloody rain that fell in parts of Britain. Also, a bloody sword was seen in the sky for three nights just after sunset. (4)

442 A.D.—At York, In England, it rained blood. (4)

445 A.D.—Apparitions were seen in the air over Britain. At Rome, burning spears were seen in the sky. This was shortly before a Saxon army was defeated by the Britains. (4)

523 A.D.—Near Kent, in England, wild beasts and dragons were seen fighting in the sky. Afterwards, it rained blood and wheat. (4)

540 A.D.—Roger of Wendover records that in this year a great comet was seen in Gaul. It was so large that the sky seemed to be on fire. (4)

580 A.D.—According to Gregory of Tours, in the 5th year of the reign of King Childebert, a great fireball was seen in the sky over Touraine, France. It was in the morning, just before sunrise, when a bright light shot across the sky disappearing in the east. This was followed by a loud sound as if trees came crashing to the ground. (5)

583 A.D.—In France, a ball of fire fell from the sky and moved a great distance through the air before it disappeared behind a cloud. (5)

589 A.D.—Gregory of Tours mentions a number of fiery globes traveling through the sky. (5)

596 A.D.—At Surrey, England strange things were seen in the sky with loud noises and flashes like lightning. (4)

655 A.D.—It is recorded in England that fire fell from heaven, and great fear came upon men. (4)

661 A.D.—A globe of fire fell onto St. Pauls Church in London and burnt the roof. (This could be a case of Ball Lightning.) (4)

675 A.D.—At a cloister and monastery in Barking, England, Bede Writes " A light from heaven like a great sheet suddenly appeared . . ." The light lingered for a moment before shooting upward and out of sight. (6)

685 A.D.—There was a bloody rain that fell in Britain. (4)

715 A.D.—In England, many strange and frightening sights appeared in the sky, including monstrous creatures and armies battling. Afterwards, a great storm caused houses to overturn and great oaks to uproot. (4)

719 A.D.—Hail burnt ships, and the sea was said to boil off the coast of Britain. (4)

735 A.D.—A great fireball was seen in England near the end of Autumn. (4)

745 A.D.—(January 01) In Britain, fiery strokes were seen in the heavens, and dragons and ships seen in the air. (4)

773 A.D.—In Britain, after sunset, a red cross appeared in the heavens. (4)

785 A.D.—At Clonmacnoise, in Ireland, a "dreadful vision" was seen in the sky. (4)

793 A.D.—Dragons and fire were seen shooting through the heavens in Britain. (4)

796 A.D.—Globes of fire were seen around the sun. (4)

810 A.D.—Einhard, in his *Vita Caroli* writes about a bright fireball that was seen during the Emperor Charlemagne's campaign into Saxony. Shortly before sunrise, the army set out on the day's march. The emperor was riding along on his horse near the front of his army when the fireball struck. Einhard says that " Charlemagne saw a meteor flash down from the heavens and pass along the clear sky from right to left with a great blaze of light." This must have caused quite a fright for his horse, for it lowered its head and fell. This caused Charlemagne to tumble to the ground. The force of the fall was so violent that it broke the buckle on his cloak, and his sword and belt were torn off. At the time he was thrown he

was holding a javelin. This weapon was found more than 20 feet away. The emperor in no way believed that this fireball was a bad omen. He merely shook it off and went on with business as usual. (7)

813 A.D.—Alfred of Beverley records meteors seen in the sky over Britain. (4)

861 A.D.—At Nogata, in Japan, a fireball blazed through the sky. A stone was later recovered by the towns residents, and is kept at the Shinto shrine of Suga Jinja. This is the oldest recorded fall of a meteorite in which the stone still exists.

867 A.D.—The 18th century historian Thomas Short mentions a cloud that was seen hanging in the sky over England which appeared to be half blood and half fire. He added that this was followed by a Danish invasion. (4)

912 A.D.—In Scotland, fiery torches were seen in the air along with 4 rainbows. (4)

917 A.D.—Burning comets were seen in the sky over Ireland.

921 A.D.—A large fireball plunged into a river at Narni, Italy. (1)

934 A.D.—In Ireland, it is said the mountains near Connaught were burnt with celestial fire. The lakes and rivers were dried up and people were burnt. (4)

945 A.D.—(October 25) In Ireland, 2 fiery columns were seen a week before Halloween. It is said that these brilliant lights "illuminated the whole world." (4)

979 A.D.—Simeon of Durham, an English monk writes about a bloody cloud that changes later to fire. (4)

991 A.D.—On Christmas day the Annals of Ulster mention a bright light seen in the sky over Ireland that resembled a burning hand. (4)

1066—A comet is seen in the sky over Britain and Normandy. This was almost certainly Halley's comet which appeared early in the year. One famous seen of the Bayeaux Tapestry shows a court page, or some type of messenger whispering into the ear of King Harold of England. Near Harold is an image that has been interpreted as a comet. There is a Latin inscription next to it which says "ISTI MIRANT STELLA." Translated, this means "They wonder at the star." In those days, comets were looked at with a degree of fear and superstition. Harold must have wondered what it meant. Only a few months after the comet appeared, Harold would lose his kingdom when a Norman invasion fleet led by Duke William crossed the English Channel and soundly defeated a tired English army at the Battle of Hastings. (9)

1067—(December 06) The sky seemed to be on fire in Northumbria at various times during the year. This was possibly due to the Aurora Borealis, or some type of Volcanic activity somewhere on the planet. (4)

1098—(September 27) Simeon of Durham reports that a star like a comet appeared in the heavens and stayed for 15 days. Other people reported seeing something like a burning cross in the sky. (4)

1131—(January 11 or 12) According to the Anglo-Saxon Chronicle, the heavens in the northern sky were on fire. (4)

1137—An Irish cleric Thady Dowling , in the Annals of Ireland, states that in this year 3 red suns appeared which brought war to the British Isles. (4)

1138—(October 07) It is recorded that in England the heavens were seen to "emit fiery sparks like a furnace, and balls of fire of wonderful brightness, like the sparks of live coals shot through the air in more places than one." (4)

1158—(August 28) C.E. Britton mentions an incident he found in the *Annals of Clonmacnoise*. In Ireland "there was a great amount of fier seen in the firmament this yeare, westerly of Tea Doynn in Munster, it was bigger then St. Patricks mount, which dispersed in severall showers of small sparkles of fier without doing any hurt, this was upon the eve of St. john in Autumne." (4)

1164—(September 18) Three circles of four different colors, like a rainbow were seen in the sky over England. When they disappeared there were two suns in the sky. (4)

1168—In Britain, a globe of fire was seen in the sky. (4)

1170—The *Annals of Boyle* mentions a bow of burning fire seen in Ireland. (4)

1173—(February 10) In Londonderry, Ireland a fireball blazed across the sky and stayed in the southeast part of the sky. People arose from their beds thinking that morning had come. Gervase of Canterbury also mentions this incident. In England he says "there appeared some time after midnight a wonderful sign in the sky. For a certain red color was seen in the air in the northerly regions between east and west. White rays were also traversing this redness, now slender like spears, now broad like tables, and now here and now there as if erected upwards from Earth to heaven. The aforesaid white rays were like beams of the sun penetrating the thickest clouds." (4)

1177—(November 29) Gervase of Canterbury records a strange event that took place in the sky over Kent, England. It took place before the first hour of the vigil of St. Andrew. A red burning flame was seen in the sky which some people took to be that of "fiery dragons with many heads." (4)

1178—(June 23) Gervase of Canterbury writes about another strange event that took place in the sky over England. He states that " it occurred the day after the nativity of John the Baptist when the moon was full." He records " from the east with the moon shining there sprung up a burning flame which threw forth sparks, the witnesses were uneasy as the moon was struck hard and slayed." Obviously, the monks who were witness to this fireball believed that it hit the moon. In reality this was almost certainly an optical illusion. The bolide was probably in the earth's atmosphere and merely ran it's course at about the same time it passed the moons position in the sky. (see chapter 4) (10)

1188—(August 09) A cross appeared in the sky over Dunstable, England which many saw the image of Jesus Christ fastened with nails. (4) (11)

1200—Roger de Hoveden states that around Christmas five moons appeared in the sky over York England. (4)

1204—(April 01) In England, a "red light , like fire was seen in the sky" This lasted until midnight. It was also noted that the stars around it appeared to be red. (4)

1217—(October 27) The *Dunstable Annals* records that on the vigil of Simon and Jude a large cross was seen in the sky "passing through the air with great glory from the Eastern to the Western parts." (4)

1218—The *Lanecrost Chronicle* states that "prodigious appearences were seen in the sky." (4)

1222—William of Newburgh records that a large comet was seen in the sky along with many dragons. The *Waverly Annals* also mentions that in this year in the sky over England, dragons were seen fighting. (4)

1233—(June) Roger of Wendover states that in Britain two snakes were seen fighting in the sky. One of them finally overcame the other. (4)

1243—(June 26) Matthew of Paris records what appears to be a meteor shower. "Stars fell from the sky 30 or 40 in a single instant." (4)

1254—(January 01) On New Years Day, Matthew of Paris records a prodigy seen in the sky by some monks over St. Albans in England. "Indeed, in the night of the Lord's circumcision, at midnight, the air being most serene, and the sky covered with stars, the moon being eight days old, there appeared in the air marvelous to relate, a kind of large ship, elegantly shaped, equipped and of marvelous color. Certain monks at St. Albans saw this appearance, being at St. Amphibalus' to commemorate the festival, and looking at the stars to see if it was yet the hour for singing mattins, and they called together all of their familiar friends who were in the house to see the marvel. It appeared for a long time as if it were painted and in truth a ship made of planks, but finally it began to disappear whence it was believed to be a cloud, but a marvelous and prodigious one." (4)

1263—(July 29) Matthew of Westminster (now known to be a pseudonym for a number of different chroniclers of the late 13th and early 14th centuries) writes that a wonderful sign appeared in the northern part of the sky over England at about midnight. (4)

1274—(December 05) Fiery dragons and a comet were seen in the sky over England (Matthew of Westminster) (4)

1282—(March 15) The *Worcester Annals* records "On the Ides of March there was seen by the prior of the order of St. Augustine at Lodelaw, with the Lord Briennus of Brompton and other knights and many others at Kinlet, three suns, one in the east, another in the west and a third in the south. (Probably Parhelion) (4)

1285—A "flaming globe" coming after a strong wind crossed over a river and destroyed two houses. This was recorded in the *Lanecrost Chronicle*. (4)

1286—(May 01) At about midnight, and lasting for a few hours, a sound was heard by many people. This event was recorded in the *Worcester Annals*. The noise was probably caused by a fireball. The longevity of the sound can be explained by the number of people who reported hearing it and the discrepencies of the hour in which it was heard. (4)

1286—(May 08) John of Everisden records that in Suffolk, England, armies were seen fighting in the air. (4)

1287—(January 14) John of Everisden records an incident from the town of Bury St. Edmunds in England. "On the morrow of the octave of the Epiphany, sudden flashes of light were seen which terrified the beholders. (4)

1291—(February 08) According to John of Everisden, a loud and sharp report was heard in London which seemed to come from the sky. (4)

1322—(November 04) Matthew of Westminster (or someone using that name) writes of an incident that took place in the sky over Uxbridge, England. It is said that "on the fourth of November at the first hour of the night in the western parts beyond the city of London near the village of Uxbridge, there appeared in the air to many beholders a wonderful sign. For a certain pile of fire of the size and shape of a small boat, pallid, but of livid color rising up from the south and crossing the firmament with a slow and grave motion, set its course towards the north. Out of the front of this pile another very fervent fire of a red color and of greater quantity, similar in shape to the former, burst forth immediately with bright beams and great speed, flying through the air, which were seen quickly meeting against each other by many beholders. And by turns frequently approaching with collisions and engaging in fearful combat, the blows of which conflict and the sounds of the crashes were heard at distances from the beholders. (4)

1327—The *Chronicles of London* state "In this yere were seyn in the firmament two mones (moons) and this yere were two popes." According to the list of popes issued by the Vatican there was only one pope in the year 1327, John XII. I assume the second pope must be a reference to the anti-pope Nicholas V. (4) (List of Popes)

1342—(October 11) The *Annals of Ireland* reports that two moons were seen in the sky over Dublin. One of these moons was bright and located in the western sky. The other was in the eastern sky and shaped like a loaf. (4)

1355—Henry Knighton, an Augustinian canon mentions an incident that occurred in the sky over England which could have been the Aurora Borealis. It was said that "two banners, one red and one blue were fighting one another until finally the red banner defeated the blue one and cast it down on the Earth." (4)

1360—The Chronicon Angliae mentions a strange prodigy seen in the sky over England and France. It says " there suddenly appeared two towers, form which two armies went out, one of which was crowned with a warlike sign, and the other was clothed in a black color. They met and the soldiers overcame those in black. A second time the warriors overcame the blacks and returned to their tower, and the whole vanished. (4)

1361—(February 25) (*Eulogium Historiarum*) A luminous cloud appeared at midnight which looked like fire. In this fire were seen images like men. (4)

1366—(October 31) John of Reading records that there were falling stars in the sky which fell and burned the clothes of people that they fell on. (4)

1385—(July 15) A fire in the shape of a head appeared over London and Dover. (John of Malvern) (4)

1387—(November and December) Mentioned in *Knighton's Continuator* "A certain appearance in the likeness of a fire was seen in many parts of the kingdom of England. "It is said that this fire sometimes took the shape of a burning wheel and at other times a round barrel or a long beam." (4)

1388—(April) A flying dragon was seen in many places in England. (*Knighton's Continuator)* (4)

1492—(October 11) Columbus sees a strange light while standing on the stern castle of the *Santa Maria.* The light was described as "like a little wax candle bobbing up and down." This light was also witnessed by Columbus' servent Pero Gutierrez. Over the centuries a number of possible explanations have been proposed. The most popular one was that the light was a torch carried in the hand of a person walking along the beach of San Salvador. The 19ᵗʰ century American writer Washington Irving who is best known for his short story *The Legend of Sleepy Hollow* suggested that the light might have been caused by "a torch in the bark of a fisherman, rising and sinking with the waves . . ." Another 19ᵗʰ century scholar and biographer of Columbus, Justin Winsor was a bit more skeptical, and adds a little arithmetic to the equation. Winsor believed that Columbus was at least 12 to 14 leagues from San Salvador when he observed the light. Given that distance and the low elevation of San Salvador it appears highly unlikely that the light he saw came from the island. The best explanation of what he saw was that it was a fireball off in the distance. It would have been quite a distance from the ship, but the illusion of it bobbing up and down could be attributed to the rolling of the ship on the waves and its proximity to the horizon. (12) (13)

1492—(November 07) A large fireball was seen over the small city of Ensisheim, France. The meteor struck between 11:00 A.M. and noon, and was seen by a number of people. It landed in a wheat field, and only one witness, a small child, was able to lead the city authorities to the place of fall. To their amazement the people of Ensisheim discovered a large crater 5 feet deep in which there was a stone that was later estimated to weigh between 200 and 260 pounds. Eventually it was removed from the hole and broken up by souvenir hunters. One of the largest pieces was given to the local church by the Hapsburg king and future Holy Roman Emperor Maximilian II. Today this stone resides in the Regency Palace in Ensisheim. (1)

1511—(September 14) In Italy, at Crema, stones fell from the sky. Some of these stones were recovered, and were said to weigh 100 pounds each. A monk was killed by a fragment from one of these stones. (1)

1516—Near the Abdua River in China it is said that 1200 stones fell from the sky. (14)

1540—(April 28) A large stone the size of a barrel fell at Limousin in France. The stone was said to have made a crater several feet deep. (14)

1622—(January 10) A meteor fell in Cornwall, England. (11)

1628—(April 09, 5:00 P.M.) At Hatford, England it was said that a "hideous noise" was heard in the sky followed by a stone which fell from the sky, and was found by a woman a mile and a half from Hatford. (11)

1637—(November 29) A stone weighing 54 pounds fell on Mt. Vaison in Provence, France. (1)

1638—(October 21) A stone fell from the sky at Wydecombe near Dartmoor, England. (11)

1642—(August 04) According to *Gentlemen's Magazine,* a British publication, in Suffolk, between Woodbridge and Albons a noise like a drum beating was heard in the sky. This was followed by a stone falling from the heavens which was found by a certain Captain Johnson who found that it was hot when he dug it out of the ground. (11)

1650—(March 30) The 19th century British astronomer Daniel Kirkwood wrote about a Franciscan monk that was killed by a falling stone in Milan, Italy. The *London Encyclopedia* mentions this same incident, but says that it occurred in the year 1654. (1)

1652—(May) Ernst Chaldni mentions a fireball that fell from the sky near Sienna and Rome. (16)

1672—(November 22) At Wednesbury, in the Black Country of England there was seen a large bright fireball which burst from the heavens. (11)

1676—(September 20) An extremely bright fireball was seen over much of England at 7:00 P.M. It was described as being "as light as noonday." (11)

1676—A stone of meteoric origins fell into a boat in the Orkney's. (11)

1677—(July 17) Increase Mather mentions an incident in his book *Remarkable Providences* that took place on a sailing vessel off the coast of Cape Cod near Massachusetts. The captain of the ship stated that he saw "a very black object fly before him about the bigness of a small mast." He also heard a sound "as if great armies of men had been firing one against the other." The blast shook the vessel, and the captain was briefly rendered unconscious. After recovering his senses he assessed the damages that the strange object had done to his ship, and found to his amazement that the main mast had been split in half, and the deck was punctured by tiny holes. There was also a strong odor of sulfur in the air. The captain was forced to turn his ship around and head toward Boston where it could be refitted. (15)

1680—Several stones fell from the sky near Gresham College in London. (11)

1686—(January 31) A strange substance like "charred paper" fell from the sky in Norway. (16)

1708—(July 31) Edmund Halley reported that a large meteor was seen over London and parts of Suffolk. (11)

1710—A large meteor was seen on holy Thursday over Leeds, Nottingham, Derby and Lancashire at 10:15 P.M. (11)

1719—(March 19) Edmund Halley reported that a large fireball was seen over much of England. It was described as being "not as large as the moon, but almost as bright as the sun, and bursting with a loud report." This fireball was also seen in Scotland and France. (11)

1725—(July 03) A stone fell from the sky near Mixbury, England. It was recovered and said to have weighed 20 pounds. (11)

1731—(March 12) A fireball was seen in the sky over Halstead, England. Supposedly, this stone fell into the water, and the water was said to "froth and Foam" for 30 hours after the fall. This incident could have happened in 1732, sources vary. (11)

1731—(December 09) At Florence, Italy, a "luminous cloud" was seen moving at rapid speed across the sky. It eventually disappeared. (16)

1733—(December 08) A kite shaped meteor was seen in the sky over Walkhampton and Oakhampton, England.

1734—(March 13) A fireball seen over London. (11)

1737—(December 05) At Kilkenny, Ireland a large fireball was seen in the sky before exploding. This was seen at about the same time as an Aurora Borealis. (11)

1738—(August 28) At 5:00 P.M. with the sun shining, a fireball was seen in the sky over many parts of England, including Dorsetshire, Devonshire, Berkshire, and Derbyshire. (11)

1741—(December 11) A meteor the size of the moon was seen over London, Canterbury, Peckham, and Hungerford at 1:00 P.M. The meteor was said by witnesess to be kite shaped, and exploded with a loud clap. (11)

1741—(December 18) A "ball of fire" was seen at Canterbury followed by a storm that broke many windows. (11)

1742—(December 16) A large meteor was seen over London. (11)

1744—(May 27) A large meteor seen over Somerset Gardens in London. (11)

1745—(July 14) A large meteor was seen over Oxford in England, and was said to have lasted for one hour and one minute, while changing into different shapes. (Who knows what to make of this one, it could have been the Aurora Borealis.) (11)

1750—(July 22) At 8:40 P.M. a "very brilliant" meteor was seen over Peterborough, and Norwich, England. (11)

1752—(April 15) A strange looking star in the shape of an octagon seen in the sky over Slavange, Norway. (16)

1755—(October 15) Meteors in the sky over Lisbon, Portugal. (16)

1777—(June 17) "Dark spherical bodies" seen in the sky. (16)

1779—(June 07) Numerous "luminous bodies and globes" seen in the sky over Boulogne, France. (16)

1786—(September 02) From *Gregs Catalogues* a "bright ball of fire and light" was witnessed by many in England during a hurricane and was said to be visible for 40 minutes. (16)

1801—(June) In the sky above Youghal, Cork County, Ireland, a type of mirage was seen in the sky. There appeared "mansions, surrounded by shrubbery and white palings." (16)

1803—(April 26) At L'Aigle, France a large fireball broke up scattering fragments over much of this small farming community.

1807—(December 14) A large fireball was seen over a large area of New England and New York. This meteor burst over the small town of Weston Connecticut leaving a trail of debris across a wide area.